Student Solutions Manual for Gallian's Contemporary Abstract Algebra

Redesigned for the 11th edition of *Contemporary Abstract Algebra, Student Solutions Manual,* written by the author, has comprehensive solutions for all odd-numbered exercises and many even-numbered exercises. This Manual also offers many alternative solutions to those appearing in the text. These will provide the student with a better understanding of the material. This is the only available student solutions manual prepared by the author of *Contemporary Abstract Algebra, Eleventh Edition* and the only official one. It is designed to supplement the text and the author's original approach to instruction.

Joseph A. Gallian earned his PhD from Notre Dame. In addition to receiving numerous national awards for his teaching and exposition, he has served terms as the Second Vice President, and the President of the MAA. He has served on 40 national committees, chairing ten of them. He has published over 100 articles and authored six books. Numerous articles about his work have appeared in the national news outlets, including the *New York Times,* the *Washington Post,* the *Boston Globe,* and *Newsweek,* among many others.

Textbooks in Mathematics

Series editors:

Al Boggess, Kenneth H. Rosen

Vector Calculus
Steven G. Krantz and Harold Parks

Intuitive Axiomatic Set Theory
José Luis García

Fundamentals of Abstract Algebra
Mark J. DeBonis

A Bridge to Higher Mathematics
James R. Kirkwood and Raina S. Robeva

Advanced Linear Algebra, Second Edition
Nicholas Loehr

Mathematical Biology: Discrete and Differential Equations
Christina Alvey and Daniel Alvey

Numerical Methods and Analysis with Mathematical Modelling
William P. Fox and Richard D. West

Business Process Analytics
Modeling, Simulation, and Design
Manuel Laguna and Johan Marklund

Quantitative Literacy Through Games and Gambling
Mark Hunacek

Measure Theory and Integral
Theory and Practice
John Srdjan Petrovic

Contemporary Abstract Algebra, Eleventh Edition
Joseph A. Gallian

Student Solutions Manual for Gallian's Contemporary Abstract Algebra, Eleventh Edition
Joseph A. Gallian

https://www.routledge.com/Textbooks-in-Mathematics/book-series/CANDHTEXBOOMTH

Student Solutions Manual for Gallian's Contemporary Abstract Algebra

Eleventh Edition

Joseph A. Gallian

CRC Press
Taylor & Francis Group
Boca Raton London New York

CRC Press is an imprint of the
Taylor & Francis Group, an **informa** business
A CHAPMAN & HALL BOOK

First edition published 2025
by CRC Press
2385 NW Executive Center Drive, Suite 320, Boca Raton FL 33431

and by CRC Press
4 Park Square, Milton Park, Abingdon, Oxon, OX14 4RN

CRC Press is an imprint of Taylor & Francis Group, LLC

ISBN: 978-1-032-77957-7 (pbk)
ISBN: 978-1-003-58969-3 (eBook)

DOI: 10.1201/9781003589693

Publisher's note: This book has been prepared from camera-ready copy provided by the authors.

CONTEMPORARY ABSTRACT ALGEBRA 11TH EDITION INSTRUCTOR SOLUTIONS MANUAL

CONTENTS

Fields

Special Topics

CHAPTER 0
Preliminaries

1. $\{1,2,3,4\}$; $\{1,3,5,7\}$; $\{1,5,7,11\}$; $\{1,3,7,9,11,13,17,19\}$; $\{1,2,3,4,6,7,8,9,11,12,13,14,16,17,18,19,21,22,23,24\}$

2. **a.** 2; 10 **b.** 4; 40 **c.** 4: 120; **d.** 1; 1050 **e.** $pq^2; p^2q^3$

3. 12, 2, 2, 10, 1, 0, 4, 5.

4. $s = -3$, $t = 2$; $s = 8$, $t = -5$

5. Let a be the least common multiple of every element of the set and b be any common multiple of every element of the set. Write $b = aq + r$ where $0 \le r < a$. Then, for any element c in the set, we have that c divides $b - aq = r$. This means that r is a common multiple of every element of the set and therefore is greater than or equal to a, which is a contradiction.

6. If $n = 0 \bmod 3$, we are done. If $n = 1 \bmod 3$, the $n + 2 = 0 \bmod 3$. If $n = 2 \bmod 3$, then $n + 1 = 0 \bmod 3$.

7. By using 0 as an exponent if necessary, we may write $a = p_1^{m_1} \cdots p_k^{m_k}$ and $b = p_1^{n_1} \cdots p_k^{n_k}$, where the p's are distinct primes and the m's and n's are nonnegative. Then $\operatorname{lcm}(a,b) = p_1^{s_1} \cdots p_k^{s_k}$, where $s_i = \max(m_i, n_i)$ and $\gcd(a,b) = p_1^{t_1} \cdots p_k^{t_k}$, where $t_i = \min(m_i, n_i)$ Then $\operatorname{lcm}(a,b) \cdot \gcd(a,b) = p_1^{m_1+n_1} \cdots p_k^{m_k+n_k} = ab$.

8. The first part follows from the Fundamental Theorem of Arithmetic; for the second part, take $a = 4$, $b = 6$, $c = 12$.

9. Write $a = nq_1 + r_1$ and $b = nq_2 + r_2$, where $0 \le r_1, r_2 < n$. We may assume that $r_1 \ge r_2$. Then $a - b = n(q_1 - q_2) + (r_1 - r_2)$, where $r_1 - r_2 \ge 0$. If $a \bmod n = b \bmod n$, then $r_1 = r_2$ and n divides $a - b$. If n divides $a - b$, then by the uniqueness of the remainder, we have $r_1 - r_2 = 0$. Thus, $r_1 = r_2$ and therefore $a \bmod n = b \bmod n$.

10. Write $as + bt = d$. Then, $a's + b't = (a/d)s + (b/d)t = 1$.

11. By Exercise 9, to prove that $(a + b) \mod n = (a' + b') \mod n$ and $(ab) \mod n = (a'b') \mod n$ it suffices to show that n divides $(a + b) - (a' + b')$ and $ab - a'b'$. Since n divides both $a - a'$ and n divides $b - b'$, it divides their difference. Because $a = a' \mod n$ and $b = b' \mod n$, there are integers s and t such that $a = a' + ns$ and $b = b' + nt$. Thus $ab = (a' + ns)(b' + nt) = a'b' + nsb' + a'nt + nsnt$. Thus, $ab - a'b'$ is divisible by n.

12. Write $d = au + bv$. Since t divides both a and b, it divides d. Write $s = mq + r$, where $0 \le r < m$. Then, $r = s - mq$ is a common multiple of both a and b so $r = 0$.

13. Suppose that there is an integer n such that $ab \mod n = 1$. Then there is an integer q such that $ab - nq = 1$. Since d divides both a and n, d also divides 1. So, $d = 1$. On the other hand, if $d = 1$, then by the corollary of Theorem 0.2, there are integers s and t such that $as + nt = 1$. Thus, modulo n, $as = 1$.

14. $7(5n + 3) - 5(7n + 4) = 1$.

15. By the GCD Theorem there are integers s and t such that $ms + nt = 1$. Then $m(sr) + n(tr) = r$.

16. It suffices to show that $(p^2 + q^2 + r^2) \bmod 3 = 0$. Notice that for any integer a not divisible by 3, $a \bmod 3$ is 1 or 2 and therefore $a^2 \bmod 3 = 1$. So, $(p^2 + q^2 + r^2) \bmod 3 = p^2 \bmod 3 + q^2 \bmod 3 + r^2 \bmod 3 = 3 \bmod 3 = 0$.

17. Let p be a prime greater than 3. By the Division Algorithm, we can write p in the form $6n + r$, where r satisfies $0 \leq r < 6$. Now observe that $6n, 6n + 2, 6n + 3$, and $6n + 4$ are not prime.

18. By properties of modular arithmetic we have
$(7^{1000}) \bmod 6 = (7 \bmod 6)^{1000} = 1^{1000} = 1$. Similarly,
$(6^{1001}) \bmod 7 = (6 \bmod 7)^{1001} = -1^{1001} \bmod 7 = -1 = 6 \bmod 7$.

19. Since st divides $a - b$, both s and t divide $a - b$. The converse is true when $\gcd(s, t) = 1$.

20. If $n = ab$, where $1 < a < n$ and $1 < b < n$, then both a and b are terms in the product $(n - 1)!$. But then, $n = ab$ is a factor of $(n - 1)!$ and therefore, $(n - 1)! = 0 \bmod n$.

21. If $\gcd(a, bc) = 1$, then there is no prime that divides both a and bc. By Euclid's Lemma and unique factorization, this means that there is no prime that divides both a and b or both a and c. Conversely, if no prime divides both a and b or both a and c, then by Euclid's Lemma, no prime divides both a and bc.

22. If one of the primes did divide $k = p_1p_2 \cdots p_n + 1$, it would also divide 1.

23. Suppose that there are only a finite number of primes $p_1, p_2, \ldots, p_n$. Then, by Exercise 22, $p_1p_2 \ldots p_n + 1$ is not divisible by any prime. This means that $p_1p_2 \ldots p_n + 1$, which is larger than any of $p_1, p_2, \ldots, p_n$, is itself prime. This contradicts the assumption that $p_1, p_2, \ldots, p_n$ is the list of all primes.

24. Let $z_1 = a + bi$ and $z_2 = c + di$. Then, $z_1z_2 = (ac - bd) + (ad + bc)$;
$|z_1| = \sqrt{a^2 + b^2}, |z_2| = \sqrt{c^2 + d^2}, |z_1z_2| = \sqrt{a^2c^2 + b^2d^2 + a^2d^2 + b^2c^2} = |z_1||z_2|$.

25. x NAND y is 1 if and only if both inputs are 0; x XNOR y is 1 if and only if both inputs are the same.

26. If $x = 1$, the output is y, else it is z.

27. Let S be a set with $n + 1$ elements and pick some a in S. By induction, S has 2^n subsets that do not contain a. But there is one-to-one correspondence between the subsets of S that do not contain a and those that do. So, there are $2 \cdot 2^n = 2^{n+1}$ subsets in all.

28. Use induction and note that $2^{n+1}3^{2n+2} - 1 = 18(2^n3^{2n}) - 1 = 18(2^n3^{3n} - 1) + 17$.

29. Consider $n = 200! + 2$. Then, 2 divides n, 3 divides $n + 1$, 4 divides $n + 2, \ldots$, and 202 divides $n + 200$.

30. Use induction on n.

31. Say $p_1p_2 \cdots p_r = q_1q_2 \cdots q_s$, where the p's and the q's are primes. By the Generalized Euclid's Lemma, p_1 divides some q_i, say q_1 (we may relabel the q's if necessary). Then $p_1 = q_1$ and $p_2 \cdots p_r = q_2 \cdots q_s$. Repeating this argument at each step we obtain $p_2 = q_2, \cdots, p_r = q_r$ and $r = s$.

32. 47. Mimic Example 17.

33. Suppose that S is a set that contains a and whenever $n \geq a$ belongs to S, then $n+1 \in S$. We must prove that S contains all integers greater than or equal to a. Let T be the set of all integers greater than a that are not in S and suppose that T is not empty. Let b be the smallest integer in T (if T has no negative integers, b exists because of the Well Ordering Principle; if T has negative integers, it can have only a finite number of them so that there is a smallest one). Then $b-1 \in S$, and therefore $b = (b-1)+1 \in S$. This contradicts our assumption that b is not in S.

34. By the Second Principle of Mathematical Induction, $f_n = f_{n-1} + f_{n-2} < 2^{n-1} + 2^{n-2} = 2^{n-2}(2+1) < 2^n$.

35. For $n = 1$, observe that $1^3 + 2^3 + 3^3 = 36$. Assume that $n^3 + (n+1)^3 + (n+2)^3 = 9m$ for some integer m. We must prove that $(n+1)^3 + (n+2)^3 + (n+3)^3$ is a multiple of 9. Using the induction hypothesis we have that $(n+1)^3 + (n+2)^3 + (n+3)^3 = 9m - n^3 + (n+3)^3 =$ $9m - n^3 + n^3 + 3 \cdot n^2 \cdot 3 + 3 \cdot n \cdot 9 + 3^3 = 9m + 9n^2 + 27n + 27 = 9(m + n^2 + 3n + 3)$.

36. You must verify the cases $n = 1$ and $n = 2$. This situation arises in cases where the arguments that the statement is true for n implies that it is true for $n+2$ is different when n is even and when n is odd.

37. The statement is true for any divisor of $8^3 - 4 = 508$.

38. One need only verify the equation for $n = 0, 1, 2, 3, 4, 5$. Alternatively, observe that $n^3 - n = n(n-1)(n+1)$.

39. Since 3736 mod 24 = 16, it would be 6 p.m.

40. 5

41. Observe that the number with the decimal representation $a_9a_8\ldots a_1a_0$ is $a_9 10^9 + a_8 10^8 + \cdots + a_1 10 + a_0$. From Exercise 9 and the fact that $a_i 10^i \bmod 9 = a_i \bmod 9$, we deduce that the check digit is $(a_9 + a_8 + \cdots + a_1 + a_0) \bmod 9$. So, substituting 0 for 9 or vice versa for any a_i does not change the value of $(a_9 + a_8 + \cdots + a_1 + a_0) \bmod 9$.

42. No.

43. For the case in which the check digit is not involved, the argument given Exercise 41 applies. Denote the money order number by $a_9a_8\ldots a_1a_0c$ where c is the check digit. For a transposition involving the check digit $c = (a_9 + a_8 + \cdots + a_0) \bmod 9$ to go undetected, we must have $a_0 = (a_9 + a_8 + \cdots + a_1 + c) \bmod 9$. Substituting for c yields $2(a_9 + a_8 + \cdots + a_0) \bmod 9 = a_0$. Then cancelling the a_0, multiplying by sides by 5, and reducing module 9, we have $10(a_9 + a_8 + \cdots + a_1) = a_9 + a_8 + \cdots + a_1 = 0$. It follows that $c = a_9 + a_8 \cdots + a_1 + a_0 = a_0$. In this case the transposition does not yield an error.

44. Observe that for any integer k between 0 and 8, $k \div 9 = .kkk\ldots$.

46. 7

47. Say that the weight for a is i. Then an error is undetected if modulo 11, $ai + b(i-1) + c(i-2) = bi + c(i-1) + a(i-2)$. This reduces to the cases where $(2a - b - c) \bmod 11 = 0$.

48. $a - a = 0$; if $a - b$ is an integer k, then $b - a$ is the integer $-k$; if $a - b$ is the integer n and $b - c$ is the integer m, then $a - c = (a - b) + (b - c)$ is the integer $n + m$. The set of equivalence classes is $\{[k] | \ 0 \leq k < 1, \ \ k \text{ is real}\}$. The equivalence classes can be represented by the real numbers in the interval $[0, 1)$. For any real number a, $[a] = \{a + k \mid \text{ where } k \text{ ranges over all integers}\}$.

49. No. $(1, 0) \in R$ and $(0, -1) \in R$ but $(1, -1) \notin R$.

50. Obviously, $a + a = 2a$ is even and $a + b$ is even implies $b + a$ is even. If $a + b$ and $b + c$ are even, then $a + c = (a + b) + (b + c) - 2b$ is also even. The equivalence classes are the set of even integers and the set of odd integers.

51. a belongs to the same subset as a. If a and b belong to the subset A and b and c belong to the subset B, then $A = B$, since the distinct subsets of P are disjoint. So, a and c belong to A.

52. Suppose that n is odd prime greater than 3 and $n + 2$ and $n + 4$ are also prime. Then $n \bmod 3 = 1$ or $n \bmod 3 = 2$. If $n \bmod 3 = 1$ then $n + 2 \bmod 3 = 0$ and so is not prime. If $n \bmod 3 = 2$ then $n + 4 \bmod 3 = 0$ and so is not prime.

CHAPTER 1

Introduction to Groups

1. Three rotations: 0°, 120°, 240°, and three reflections across lines from vertices to midpoints of opposite sides.

2. Let $R = R_{120}$, $R^2 = R_{240}$, F be a reflection across a vertical axis, $F' = RF$, and $F'' = R^2F$

	R_0	R	R^2	F	F'	F''
R_0	R_0	R	R^2	F	F'	F''
R	R	R^2	R_0	F'	F''	F
R^2	R^2	R_0	R	F''	F	F'
F	F	F''	F'	R_0	R^2	R
F'	F'	F	F''	R	R_0	R^2
F''	F''	F'	F	R^2	R	R_0

3. **a.** V **b.** R_{270} **c.** R_0 **d.** $R_0, R_{180}, H, V, D, D'$ **e.** none

4. Five rotations: 0°, 72°, 144°, 216°, 288°, and five reflections across lines from vertices to midpoints of opposite sides.

5. D_n has n rotations of the form $k(360°/n)$, where $k = 0, \ldots, n-1$. In addition, D_n has n reflections. When n is odd, the axes of reflection are the lines from the vertices to the midpoints of the opposite sides. When n is even, half of the axes of reflection are obtained by joining opposite vertices; the other half, by joining midpoints of opposite sides.

6. A nonidentity rotation leaves only one point fixed – the center of rotation. A reflection leaves the axis of reflection fixed. A reflection followed by a different reflection would leave only one point fixed (the intersection of the two axes of reflection), so it must be a rotation.

7. A rotation followed by a rotation either fixes every point (and so is the identity) or fixes only the center of rotation. However, a reflection fixes a line.

8. In either case, the set of points fixed is some axis of reflection.

9. Observe that $1 \cdot 1 = 1$; $1(-1) = -1$; $(-1)1 = -1$; $(-1)(-1) = 1$. These relationships also hold when 1 is replaced by a "rotation" and -1 is replaced by a "reflection."

10. Reflection.

11. Thinking geometrically and observing that even powers of elements of a dihedral group do not change orientation, we note that each of a, b and c appears an even number of times in the expression. So, there is no change in orientation. Thus, the expression is a rotation. Alternatively, as in Exercise 9, we associate each of a, b and c with 1 if they are rotations and -1 if they are reflections and we observe that in the product $a^2b^4ac^5a^3c$ the terms involving a represent six 1s or six -1s, the term b^4 represents four 1s or four -1s, and the terms involving c represent six 1s or six -1s. Thus the product of all the 1s and -1s is 1. So the expression is a rotation.

12. n is even.

13. In D_4, $HD = DV$ but $H \neq V$.

14. D_n is not commutative.

15. R_0, R_{180}, H, V

16. Rotations of 0° and 180°; Rotations of 0° and 180° and reflections about the diagonals.

17. R_0, R_{180}, H, V

18. Let the distance from a point on one H to the corresponding point on an adjacent H be one unit. Then, a translations of any number of units to the right or left are symmetries; a reflection across the horizontal axis through the middle of the H's is a symmetry; and a reflection across any vertical axis midway between two H's or bisecting any H is a symmetry. All other symmetries are compositions of finitely many of those already described. The group is non-Abelian.

19. In each case the group is D_6.

20. D_{28}

21. First observe that $X^2 \neq R_0$. Since R_0 and R_{180} are the only elements in D_4 that are squares we have $X^2 = R_{180}$. Solving $X^2Y = R_{90}$ for Y gives $Y = R_{270}$.

22. Their only symmetry is the identity.

23. The n rotations of D_n are $R_0, R_{360/n}, R^2_{360/n}, \ldots, R^{n-1}_{360/n}$. Suppose that $n = 2k$ for some positive integer k. Then $R^k_{360/n} = R_{360k/2k} = R_{180}$. Conversely, if $R^k_{360/n} = R_{180}$ then $360k/n = 180$ and therefore $2k = n$.

24. $X^2 = F$ has no solutions; the only solution to $X^3 = F$ is F.

25. Let p be any vertex on the n-gon and p' be the counterclockwise vertex adjacent to p. Let F be the reflection that passes through p and let F' be the reflection that passes through the midpoint point of the edge joining p and p'. Then, in the composition FF', then reflection F fixes p and the reflection F' takes p to p'. So, $FF' = R_{360}/n$.

26. Z_4, D_5, D_4, Z_2
D_4, Z_3, D_3, D_{16}
D_7, D_4, D_5, Z_{10}

CHAPTER 2

Groups

1. **c, d**

2. **c, d**

3. none

4. **a, c**

5. 7; 13; $n-1$; $\frac{1}{3-2i}=\frac{1}{3-2i}\frac{3+2i}{3+2i}=\frac{3}{13}+\frac{2}{13}i$

6. **a.** $-31-i$ **b.** 5 **c.** $\frac{1}{12}\begin{bmatrix} 2 & -3 \\ -8 & 6 \end{bmatrix}$ **d.** $\begin{bmatrix} 2 & 4 \\ 4 & 6 \end{bmatrix}$.

7. Let $A=\begin{bmatrix} 2 & 0 \\ 0 & 1 \end{bmatrix}$. Then $A \in G_1$ and $\det A = 2$ but $\det A^2 = 0$. So G_1 is not closed under multiplication. Also $A \in G_2$ but $A^{-1}=\begin{bmatrix} 1/2 & 0 \\ 0 & 1 \end{bmatrix}$ is not in G_2. G_3 is a group.

8. Say, x is the identity. Then, $0-x=0$. So, $x=0$. But $0-1\neq 1$.

9. If $5x=3$ multiply both sides by 4, we get $0=12$. If $3x=5$ multiply both sides by 7, we get $x=15$. Checking, we see that $3\cdot 15 = 5 \bmod 20$.

10. $1,3,7,9,11,13,17,19.1,9,11$, and 19 are their own inverses; 3 and 7 are inverses of each other as are 11 and 13.

11. One is Socks-Shoes-Boots.

12. The set does not contain the identity; closure fails.

13. Under multiplication modulo 4, 2 does not have an inverse. Under multiplication modulo 5, $\{1,2,3,4\}$ is closed, 1 is the identity, 1 and 4 are their own inverses, and 2 and 3 are inverses of each other. Modulo multiplication is associative.

14. $\begin{bmatrix} 1 & 1 \\ 0 & 1 \end{bmatrix}\begin{bmatrix} 1 & 0 \\ 1 & 1 \end{bmatrix} \neq \begin{bmatrix} 1 & 0 \\ 1 & 1 \end{bmatrix}\begin{bmatrix} 1 & 1 \\ 0 & 1 \end{bmatrix}$.

15. a^{11}, a^6, a^4, a^1

16. The identity is 25.

17. (a) $2a+3b$; (b) $-2a+2(-b+c)$; (c) $-3(a+2b)+2c=0$

18. $(ab)^3 = ababab$ and $(ab^{-2}c)^{-2}=((ab^{-2}c)^{-1})^2=(c^{-1}b^2a^{-1})^2=c^{-1}b^2a^{-1}c^{-1}b^2a^{-1}$.

19. Observe that $a^5=e$ implies that $a^{-2}=a^3$ and $b^7=e$ implies that $b^{14}=e$ and therefore $b^{-11}=b^3$. Thus, $a^{-2}b^{-11}=a^3b^3$. Moreover, $(a^2b^4)^{-2}=((a^2b^4)^{-1})^2=(b^{-4}a^{-2})^2=(b^3a^3)^2$.

20. $K=\{R_0,R_{180}\}$; $L=\{R_0,R_{180},H,V,D,D'\}$.

21. The set is closed because det $(AB) = (\det A)(\det B)$. Matrix multiplication is associative. $\begin{bmatrix} 1 & 0 \\ 0 & 1 \end{bmatrix}$ is the identity. Since $\begin{bmatrix} a & b \\ c & d \end{bmatrix}^{-1} = \begin{bmatrix} d & -b \\ -c & a \end{bmatrix}$ its determinant is $ad - bc = 1$.

22. $1^2 = (n-1)^2 = 1$.

23. Using closure and trial and error, we discover that $9 \cdot 74 = 29$ and 29 is not on the list.

24. All we need do is find an x with the property $xab = bax$. The solution is $x = b$.

25. For $n \geq 0$, we use induction. The case that $n = 0$ is trivial. Then note that $(ab)^{n+1} = (ab)^n ab = a^n b^n ab = a^{n+1}b^{n+1}$. For $n < 0$, note that $e = (ab)^0 = (ab)^n(ab)^{-n} = (ab)^n a^{-n}b^{-n}$ so that $a^n b^n = (ab)^n$. In a non-Abelian group $(ab)^n$ need not equal $a^n b^n$.

26. The "inverse" of putting on your socks and then putting on your shoes, is taking off your shoes then taking off your socks. Use D_4 for the examples. (An appropriate name for the property $(abc)^{-1} = c^{-1}b^{-1}a^{-1}$ is "Socks-Shoes-Boots Property.")

27. Suppose that G is Abelian. Then by Exercise 26, $(ab)^{-1} = b^{-1}a^{-1} = a^{-1}b^{-1}$. If $(ab)^{-1} = a^{-1}b^{-1}$ then by Exercise 24 $e = aba^{-1}b^{-1}$. Multiplying both sides on the right by ba yields $ba = ab$.

28. By definition, $a^{-1}(a^{-1})^{-1} = e$. Now multiply on the left by a.

29. The case where $n = 0$ is trivial. For $n > 0$, note that $(a^{-1}ba)^n = (a^{-1}ba)(a^{-1}ba)\cdots(a^{-1}ba)$ (n terms). So, cancelling the consecutive a and a^{-1} terms gives $a^{-1}b^n a$. For $n < 0$, note that $e = (a^{-1}ba)^n(a^{-1}ba)^{-n} = (a^{-1}ba)^n(a^{-1}b^{-n}a)$ and solve for $(a^{-1}ba)^n$.

30. $(a_1a_2\cdots a_n)(a_n^{-1}a_{n-1}^{-1}\cdots a_2^{-1}a_1^{-1}) = e$

31. By closure we have $\{1, 3, 5, 9, 13, 15, 19, 23, 25, 27, 39, 45\}$.

32. $f(x) = x$ for all x. See Theorem 0.8.

33. Suppose x appears in a row labeled with a twice. Say $x = ab$ and $x = ac$. Then cancellation gives $b = c$. But we use distinct elements to label the columns.

34. Z_{105}; $Z_{40}, D_{20}, U(41)$

35. Closure and associativity follow from the definition of multiplication; $a = b = c = 0$ gives the identity; we may find inverses by solving the equations $a + a' = 0$, $b' + ac' + b = 0$, $c' + c = 0$ for a', b', c'.

36. $(ab)^2 = a^2b^2 \Leftrightarrow abab = aabb \Leftrightarrow ba = ab$.
$(ab)^{-2} = b^{-2}a^{-2} \Leftrightarrow b^{-1}a^{-1}b^{-1}a^{-1} = b^{-1}b^{-1}a^{-1}a^{-1} \Leftrightarrow a^{-1}b^{-1} = b^{-1}a^{-1} \Leftrightarrow ba = ab$.

37. Since e is one solution, it suffices to show that nonidentity solutions come in distinct pairs. To this end, note that if $x^n = e$ and $x \neq e$, then $(x^{-1})^n = e$ and $x \neq x^{-1}$. So if we can find one nonidentity solution we can find a second one. Now suppose that a and a^{-1} are nonidentity elements that satisfy $x^n = e$ and b is a nonidentity element such that $b \neq a$ and $b \neq a^{-1}$ and $b^n = e$. Then, as before, $(b^{-1})^n = e$ and $b \neq b^{-1}$. Moreover, $b^{-1} \neq a$ and $b^{-1} \neq a^{-1}$. Thus, finding a third nonidentity solution gives a fourth one. Continuing in this fashion, we see that we always have an even number of nonidentity solutions to the equation $x^n = e$.

38. Note that $(\frac{1}{2}, \frac{1}{3}) = (\frac{2}{4}, \frac{1}{3})$, but $(\frac{1}{2}, \frac{1}{3})$ corresponds to $\frac{2}{5}$ whereas $(\frac{2}{4}, \frac{1}{3})$ corresponds to $\frac{3}{7}$. So, the correspondence is not a function from $Q^+ \times Q^+$ to Q^+.

39. For D_6 the subgroups of order 6 have the form $\{R_0, R_{60}, R_{120}, R_{180}, R_{240}, R_{300}\}$ and $\{R_0, R_{120}, R_{240}, F, R_{120}F, R_{240}F\}$ where F is a reflection. For the general case where $n > 3$ is even, we have the set of n rotations is a subgroup of order n. For a second subgroup of order n, let $R = R_{360/n}$ and F be any reflection in D_n. Then $H = \{R_0, R^2, \ldots, R^{n-2}, F, R^2F, \ldots, R^{n-2}F\}$ is a subgroup of D_n of order n. (Closure follows from the relation $FR^i = R^{-i}F$.) Since H contains only half of the reflections in D_n, we can obtain a third subgroup of order n by replacing F in H by any reflection F' not in H. To see that these three subgroups are the only ones, note that any other subgroup K in D_n $(n \geq 4)$ must have the same $n/2$ rotations and $n/2$ reflections. But every reflection was used in the two subgroups described above, so K would be contained in one of the two. So, the number of subgroups of order $n \geq 4$ when n is even is 3.

40. Observe that $F_1F_2 = F_2F_1$ implies that $(F_1F_2)(F_1F_2) = R_0$. Since F_1 and F_2 are distinct and F_1F_2 is a rotation it must be R_{180}.
Alternate proof. Observe that $(F_1F_2)^{-1} = F_2^{-1}F_1^{-1} = F_2F_1 = F_1F_2$ implies that (F_1F_2) is its own inverse. Since F_1 and F_2 are distinct and F_1F_2 is a rotation it must be R_{180}.

41. Since FR^k is a reflection we have $(FR^k)(FR^k) = R_0$. Multiplying on the left by F gives $R^kFR^k = F$.

42. Since FR^k is a reflection, we have $(FR^k)(FR^k) = R_0$. Multiplying on the right by R^{-k} gives $FR^kF = R^{-k}$. If D_n were Abelian, then $FR_{360^\circ/n}F = R_{360^\circ/n}$. But $(R_{360^\circ/n})^{-1} = R_{360^\circ(n-1)/n} \neq R_{360^\circ/n}$ when $n \geq 3$.

43. Using Exercise 42 we obtain the solutions R and $R^{-1}F$.

44. $R_{\beta-\alpha}$; $R_{\alpha-\beta}$

45. Since $a^2 = b^2 = (ab)^2 = e$, we have $aabb = abab$. Now cancel on left and right.

46. If a satisfies $x^5 = e$ and $a \neq e$, then so does a^2, a^3, a^4. Now, using cancellation we have that a^2, a^3, a^4 are not the identity and are distinct from each other and distinct from a. If these are all of the nonidentity solutions of $x^5 = e$, we are done. If b is another solution that is not a power of a, then by the same argument b, b^2, b^3 and b^4 are four distinct nonidentity solutions. We must further show that b^2, b^3 and b^4 are distinct from a, a^2, a^3, a^4. If $b^2 = a^i$ for some i, then cubing both sides we have $b = b^6 = a^{3i}$, which is a contradiction. A similar argument applies to b^3 and b^4. Continuing in this fashion, we have that the number of nonidentity solutions to $x^5 = e$ is a multiple of 4. In the general case, the number of solutions is a multiple of 4 or is infinite.

47. The matrix $\begin{bmatrix} a & b \\ c & d \end{bmatrix}$ is in GL$(2, Z_2)$ if and only if $ad \neq bc$. This happens when a and d are 1 and at least 1 of b and c is 0 and when b and c are 1 and at least 1 of a and d is 0. So, the elements are

$$\begin{bmatrix} 1 & 0 \\ 0 & 1 \end{bmatrix} \begin{bmatrix} 1 & 1 \\ 0 & 1 \end{bmatrix} \begin{bmatrix} 1 & 0 \\ 1 & 1 \end{bmatrix} \begin{bmatrix} 1 & 1 \\ 1 & 0 \end{bmatrix} \begin{bmatrix} 0 & 1 \\ 1 & 1 \end{bmatrix} \begin{bmatrix} 0 & 1 \\ 1 & 0 \end{bmatrix}.$$

$\begin{bmatrix} 1 & 1 \\ 0 & 1 \end{bmatrix}$ and $\begin{bmatrix} 1 & 0 \\ 1 & 1 \end{bmatrix}$ do not commute.

48. If n is not prime, we can write $n = ab$, where $1 < a < n$ and $1 < b < n$. Then, a and b belong to the set $\{1, 2, \ldots, n-1\}$, but $0 = ab \mod n$ does not. If n is prime, let c be any element in the set. Then by the Corollary of Theorem 0.2 there are integers s and t such that $cs + nt = 1$. So, mod n we have $cs = 1$.

49. Proceed as follows. By definition of the identity, we may complete the first row and column. Then complete row 3 and column 5 by using Exercise 33. In row 2 only c and d remain to be used. We cannot use d in position 3 in row 2 because there would then be two d's in column 3. This observation allows us to complete row 2. Then rows 3 and 4 may be completed by inserting the unused two elements. Finally, we complete the bottom row by inserting the unused column elements.

CHAPTER 3

Finite Groups; Subgroups

1. $|Z_{12}| = 12; |U(10)| = 4; |U(12)| = 4; |U(20)| = 8; |D_4| = 8.$
In Z_{12},
$|0| = 1; |1| = |5| = |7| = |11| = 12; |2| = |10| = 6; |3| = |9| = 4; |4| = |8| = 3; |6| = 2.$
In $U(10)$, $|1| = 1$; $|3| = |7| = 4$; $|9| = 2$.
In $U(20)$, $|1| = 1$; $|3| = |7| = |13| = |17| = 4$; $|9| = |11| = |19| = 2$.
In D_4, $|R_0| = 1$; $|R_{90}| = |R_{270}| = 4$; $|R_{180}| = |H| = |V| = |D| = |D'| = 2$.
In each case, notice that the order of the element divides the order of the group.

2. In Q, $\langle 1/2\rangle = \{n(1/2) | \; n \in Z\} = \{0, \pm 1/2, \pm 1, \pm 3/2, \ldots\}$. In Q^*, $\langle 1/2\rangle = \{(1/2)^n | \; n \in Z\} = \{1, 1/2, 1/4, 1/8, \ldots; 2, 4, 8, \ldots\}$.

3. In Q, $|0| = 1$. All other elements have infinite order since $x + x + \cdots + x = 0$ only when $x = 0$.

4. Observe that $a^n = e$ if and only if $(a^n)^{-1} = e^{-1} = e$ and $(a^n)^{-1} = (a^{-1})^n$. The infinite case follows from the infinite case.
Alternate solution. Suppose $|a| = n$ and $|a^{-1}| = k$. Then $(a^{-1})^n = (a^n)^{-1} = e^{-1} = e$. So $k \leq n$. Now reverse the roles of a and a^{-1} to obtain $n \leq k$. The infinite case follows from the finite case.

5. By the corollary of Theorem 0.2 there are integers s and t so that $1 = ms + nt$. Then $a^1 = a^{ms+nt} = a^{ms}a^{nt} = (a^m)^s(a^n)^t = (a^t)^n$.

6. In Z, the set of positive integers. In Q, the set of numbers greater than 1.

7. In Z_{30}, $2 + 28 = 0$ and $8 + 22 = 0$. So, 2 and 28 are inverses of each other and 8 and 22 are inverses of each other. In $U(15)$, $2 \cdot 8 = 1$ and $7 \cdot 13 = 1$. So, 2 and 8 are inverses of each other and 7 and 13 are inverses of each other.

8. a. $|6| = 2, |2| = 6, |8| = 3$; b. $|3| = 4, |8| = 5, |11| = 12$; c. $|5| = 12, |4| = 3, |9| = 4$. In each case $|a + b|$ divides $\text{lcm}(|a|, |b|)$.

9. $(a^4c^{-2}b^4)^{-1} = b^{-4}c^2a^{-4} = b^3c^2a^2$.

10. $aba^2 = a(ba)a = a(a^2b)a = a^3(ba) = a^5b$.

11. For F any reflection in D_6, $\{R_0, R_{120}, R_{240}, F, R_{120}F, R_{240}F\}$.

12. Let $g \in G$, $g \neq e$. If $|g| = pm$, then $|g^m| = p$.

13. If a subgroup of D_4 contains R_{270} and a reflection F, then it also contains the six other elements $R_0, (R_{270})^2 = R_{180}, (R_{270})^3 = R_{90}, R_{270}F, R_{180}F$ and $R_{90}F$. If a subgroup of D_4 contains H and D, then it also contains $HD = R_{90}$ and $DH = R_{270}$. But this implies that the subgroup contains every element of D_4. If it contains H and V, then it contains $HV = R_{180}$ and R_0.

14. $\{R_0, R_{90}, R_{180}, R_{270}\}$, $\{R_0, R_{180}, H, V\}$, and $\{R_0, R_{180}, D, D'\}$.

15. If n is a positive integer, the real solutions of $x^n = 1$ are 1 when n is odd and ± 1 when n is even. So, the only elements of finite order in R^* are ± 1.

16. 1 or 2.

17. By Exercise 29 of Chapter 2 we have $e = (xax^{-1})^n = xa^nx^{-1}$ if and only if $a^n = e$.

18. By Exercise 17, for every x in G $|xax^{-1}| = |a|$, so that $xax^{-1} = a$ or $xa = ax$.

19. Suppose $G = H \cup K$. Pick $h \in H$ with $h \notin K$. Pick $k \in K$ with $k \notin H$. Then, $hk \in G$ but $hk \notin H$ and $hk \notin K$. $U(8) = \{1,3\} \cup \{1,5\} \cup \{1,7\}$.

20. By the corollary of Theorem 0.2 we can write $1 = 2s + nt$. Then $a^1 = a^{2s+nt} = (a^2)^s(a^n)^t = (b^2)(b^n)^t = b^{2s+nt} = b^1$.

21. $U_4(20) = \{1,9,13,17\}$; $U_5(20) = \{1,11\}$; $U_5(30) = \{1,11\}$; $U_{10}(30) = \{1,11\}$. To prove that $U_k(n)$ is a subgroup it suffices to show that it is closed. Suppose that a and b belong to $U_k(n)$. We must show that in $U(n)$, $ab \bmod k = 1$. That is, $(ab \bmod n) \bmod k = 1$. Let $n = kt$ and $ab = qn + r$ where $0 \le r < n$. Then $(ab \bmod n) \bmod k = r \bmod k = (ab - qn) \bmod k = (ab - qkt) \bmod k = ab \bmod k = (a \bmod k)(b \bmod k) = 1 \cdot 1 = 1$. H is not a subgroup because $7 \in H$ but $7 \cdot 7 = 9$ is not 1 mod 3.

22. The possibilities are 1, 2, 3, and 6. 5 is not possible, for if $a^5 = e$, then $e = a^6 = aa^5 = a$. 4 is not possible for if $a^4 = e$, then $e = a^6 = a^2a^4 = a^2$.

23. Suppose that $m < n$ and $a^m = a^n$. Then $e = a^na^{-m} = a^{n-m}$. This contradicts the assumption that a has infinite order.

24. Observe that $(ab)^n = e$ if and only if $a^{-1}(ab)^na = a^{-1}ea = e$ and that $a^{-1}(ab)^na = (ba)^n$.
Alternative solution. From Exercise 17, we have $|ab| = |a^{-1}(ab)a| = |ba|$. For the second part, we have $|abab| = |(aba)b| = |b(aba)| = |baba|$. For the last part, observe that in D_4, $|R_{90}HR_{90}| = 2$, whereas $|HR_{90}H| = 4$.

25. $\det A = \pm 1$

26. $k = 4n - 1$

27. $\langle 3 \rangle = \{3, 3^2, 3^3, 3^4, 3^5, 3^6\} = \{3,9,13,11,5,1\} = U(14)$. $\langle 5 \rangle = \{5, 5^2, 5^3, 5^4, 5^5, 5^6\} = \{5,11,13,9,3,1\} = U(14)$. $\langle 11 \rangle = \{11,9,1\} \neq U(14)$. Since $|U(20)| = 8$, for $U(20) = \langle k \rangle$ for some k it must be the case that $|k| = 8$. But $1^1 = 1$, $3^4 = 1$, $7^4 = 1$, $9^2 = 1$, $11^2 = 1$, $13^4 = 1$, $17^4 = 1$, and $19^2 = 1$. So, the maximum order of any element is 4.

28. Let A be the subset of even members of Z_n and B the subset of odd members of Z_n. If $x \in B$, then $x + A = \{x + a | a \in A\} \subseteq B$, so $|A| \le |B|$. Also, $x + B = \{x + b | b \in B\} \subseteq A$, so $|B| \le |A|$.

29. By Exercise 30, either every element of H is even or exactly half are even. Since H has odd order the latter cannot occur.

30. Suppose that K is a subgroup of D_n that has at least one reflection F. Denote the rotations of K by $R_1, R_2, \ldots, R_m$. Then $R_1F, R_2F, \ldots, R_mF$ are distinct reflections in K. If F' is any reflection in K, then $F'F = R_i$ for some i. But then $F' = R_iF$. Thus, K has exactly m reflections.

31. By Exercise 30, either every element of H is a rotation or exactly half are rotations. Since H has odd order the latter cannot occur.

32. Suppose that a and b are two elements of order 2 that commute. Then, $\{e, a, b, ab\}$ is closed and therefore a subgroup.

33. Observe that by Exercise 32 we have that for any reflection F in D_n the set $\{R_0, R_{180}, F, R_{180}F\}$ is a subgroup of order 4.

34. $\langle 2 \rangle$

35. First observe that because $6 = 30 + 30 - 54$ belongs to H, we know $\langle 6 \rangle$ is a subgroup of H. Let n be the smallest positive integer in H. Then the possibilities for n are $1, 2, 3, 4, 5$, and 6. Because $\langle 6 \rangle$, $\langle 3 \rangle$ and $\langle 2 \rangle$ contain $12, 30$ and 54, these cannot be excluded. We can exclude 1 because $\langle 1 \rangle = Z$. The same is true for 5 because $6 - 5 = 1$. Finally, if 4 is in H, then so is $6 - 4 = 2$. So, our list is complete.

36. By the corollary to Theorem 0.2, $H = Z$.

37. Suppose that H is a subgroup of D_3 of order 4. Since D_3 has only two elements of order 2, H must contain R_{120} or R_{240}. By closure, it follows that H must contain R_0, R_{120}, and R_{240} as well as some reflection F. But then H must also contain the reflection $R_{120}F$.

38. $H \cap K \neq \emptyset$, since $e \in H \cap K$. Now suppose that $x, y \in H \cap K$. Then, since H and K are subgroups, we know $xy^{-1} \in H$ and $xy^{-1} \in K$. That is, $xy^{-1} \in H \cap K$.

39. For D_6 the subgroups of order 6 have the form $\{R_0, R_{60}, R_{120}, R_{180}, R_{240}, R_{300}\}$ and $\{R_0, R_{120}, R_{240}, F, R_{120}F, R_{240}F\}$ where F is a reflection. For the general case where $n > 3$ is even, we have the set of n rotations is a subgroup of order n. For a second subgroup of order n, let $R = R_{360/n}$ and F be any reflection in D_n. Then $H = \{R_0, R^2, \ldots, R^{n-2}, R_0, F, R^2F, \ldots, R^{n-2}F\}$ is a subgroup of D_n of order n. (Closure follows from the relation $FR^i = R^{-i}F$.) Since H contains only half of the reflections in D_n, we can obtain a third subgroup of order n by replacing F in H by any reflection F' not in H. To see that these three subgroups are the only ones, note that any other subgroup K in D_n $(n \geq 4)$ must have the same $n/2$ rotations and $n/2$ reflections. But every reflection was use of in the two subgroups described above, so K would be contained in one of the two. So, the number of subgroups of order $n > 3$ is even is 3.

40. $\{R_0, R_{180}\}$

41. If $x \in Z(G)$, then $x \in C(a)$ for all a, so $x \in \bigcap_{a \in G} C(a)$. If $x \in \bigcap_{a \in G} C(a)$, then $xa = ax$ for all a in G, so $x \in Z(G)$.

42. Suppose $x \in C(a)$. Then, $xa = ax$. So $a^{-1}(xa) = a^{-1}(ax) = x$. Thus, $(a^{-1}x)a = x$, and therefore $a^{-1}x = xa^{-1}$. This shows $x \in C(a^{-1})$. The other half follows by symmetry.

43. We proceed by induction. The case that $k = 0$ is trivial. Let $x \in C(a)$. If k is positive, then by induction on k, $xa^{k+1} = xaa^k = axa^k = aa^kx = a^{k+1}x$. Since $x \in C(a)$ implies that that x commutes with a^k, we have $a^k \in C(x)$. But then $a^{-k} = (a^k)^{-1} \in C(x)$. The statement "If for some integer k, x commutes a^k, then x commutes with a" is false as can be seen in the group D_4 with $x = H$, $a = R_{90}$ and $k = 2$.

44. If $a^2 = b^2$ and $a^3 = b^3$, then $a^2a = b^2b$. Now cancel a^2 and b^2.

45. Since S is nonempty, so is $\langle S\rangle$. If $a = a_1^{m_1}a_2^{m_2}\cdots a_r^{n_r}$ and $b = b_1^{n_1}b_2^{n_2}\cdots b_t^{n_t}$, belong to $\langle S\rangle$, then $ab^{-1} = a_1^{m_1}a_2^{m_2}\cdots a_r^{m_r}b_t^{-n_t}\cdots b_2^{-n_2}b_1^{-n_1}$ belongs to $\langle S\rangle$. So, $\langle S\rangle$ is a subgroup of G that contains S. If K is any subgroup of G that contains S, then K also contains $\langle S\rangle$ by closure. Thus, H contains $\langle S\rangle$. Moreover, because $\langle S\rangle$ is a subgroup of G that contains S, it is one of the terms in the intersection. So, H is contained in $\langle S\rangle$.

46. **a.** $\langle 2\rangle$ **b.** $\langle 1\rangle$ **c.** $\langle 3\rangle$ **d.** $\langle \gcd(m,n)\rangle$ **e.** $\langle 3\rangle$.

47. Since $ea = ae, C(a) \neq \emptyset$. Suppose that x and y are in $C(a)$. Then $xa = ax$ and $ya = ay$. Thus,

$$(xy)a = x(ya) = x(ay) = (xa)y = (ax)y = a(xy)$$

and therefore $xy \in C(a)$. Starting with $xa = ax$, we multiply both sides by x^{-1} on the right and left to obtain $x^{-1}xax^{-1} = x^{-1}axx^{-1}$ and so $ax^{-1} = x^{-1}a$. This proves that $x^{-1} \in C(a)$. By the Two-Step Subgroup Test, $C(a)$ is a subgroup of G.

48. Mimic the proof of Theorem 3.5.

49. No. In D_4, $C(R_{180}) = D_4$. Yes. Elements in the center commute with all elements.

50. That $C(a) \subseteq C(a^3)$ is easy. To prove the other inclusion, observe that $a^6 = a$ so if $x \in C(a^3)$, then

$$\begin{aligned} xa = xa^6 &= x(a^3a^3) = (xa^3)a^3 = (a^3x)a^3 = a^3(xa^3) \\ &= a^3(a^3x) = (a^3a^3)x = a^6x = ax. \end{aligned}$$

For the second part of the exercise try D_6.

51. Let $H = \{x \in G|\ x^n = e\}$. Since $e^1 = e, H \neq \emptyset$. Now let $a, b \in H$. Then $a^n = e$ and $b^n = e$. So, $(ab)^n = a^nb^n = ee = e$ and therefore $ab \in H$. Starting with $a^n = e$ and taking the inverse of both sides, we get $(a^n)^{-1} = e^{-1}$. This simplifies to $(a^{-1})^n = e$. Thus, $a^{-1} \in H$. By the Two-Step test, H is a subgroup of G. In D_4, $\{x|\ x^2 = e\} = \{R_0, R_{180}, H, V, D, D'\}$. This set is not closed because $HD = R_{90}$.

52. For any integer $n \geq 3$, observe that the rotation $R_{360/n}$ in D_n has order n. Now in D_n let F be any reflection. Then $F' = R_{360/n}F$ is a reflection in D_n. Also $|F'| = |F| = 2$ and $F'F = R_{360/n}$ has order n.

53. Induction shows that for any positive integer n we have

$$\begin{bmatrix} 1 & 1 \\ 0 & 1 \end{bmatrix}^n = \begin{bmatrix} 1 & n \\ 0 & 1 \end{bmatrix}.$$

So, when the entries are from $\mathbf{R}$, $\begin{bmatrix} 1 & 1 \\ 0 & 1 \end{bmatrix}$ has infinite order. When the entries are from Z_p, the order is p.

54. $|A| = 2$, $|B| = 2$, $|AB| = \infty$.

55. First observe that $(a^d)^{n/d} = a^n = e$, so $|a^d|$ is at most n/d. Moreover, there is no positive integer $t < n/d$ such that $(a^d)^t = a^{dt} = e$, for otherwise $|a| \neq n$.

56. In Exercise 54, let $a = A^{-1}$, $b = AB$. Then, $ab = A^{-1}AB = B$ has finite order.

57. Let G be a group of even order. Observe that for each element x of order greater than 2, x and x^{-1} are distinct elements of the same order. So, because elements of order greater than 2 come in pairs, there is an even number of elements of order greater than 2 (possibly 0). This means that the number of elements of order 1 or 2 is even. Since the identity is the unique element of order 1, it follows that the number of order 2 is odd.

58. For the first part use induction; 6, ∞.

59. For any positive integer n, a rotation of $360°/n$ has order n. If we let R be a rotation of $\sqrt{2}$ degrees then R^n is a rotation of $\sqrt{2}n$ degrees. This is never a multiple of $360°$, for if $\sqrt{2}n = 360k$ then $\sqrt{2} = 360k/n$, which is rational. So, R has infinite order.

60. Let $a = 2$ and $b = -2^{-1}$.

61. Inscribe a regular n-gon in a circle. Then every element of D_n is a symmetry of the circle.

62. Let F and F' be reflections that intersect in an angle of $\sqrt{2}$ degrees. Then FF' is a rotation of $2\sqrt{2}$ degrees, which has infinite order.

63. Let $|g| = m$ and write $m = nq + r$ where $0 \leq r < n$. Then $g^r = g^{m-nq} = g^m(g^n)^{-q}$ belongs to H. So, $r = 0$.

64. **a.** 2, 2, 4 **b.** 4, 6, 24 **c.** 2, 4, 8 **d.** 2, 4, 8.

65. $1 \in H$, so $H \neq \emptyset$. Let $a, b \in H$. Then $(ab^{-1})^2 = a^2(b^2)^{-1}$, which is the product of two rationals. The integer 2 can be replaced by any positive integer.

66. $\{1, 9, 11, 19\}$.

67. Let $|a| = n$ and write $m = nq + r$ where $0 \leq r < n$. Then $e = a^m = a^{nq+r} = (a^n)^q a^r = a^r$. But that forces $r = 0$.

68. Since $|e| = 1$, H is nonempty. Assume $a, b \in H$ and Let $|a| = m$ and $|b| = n$. Then, $(ab)^{mn} = (a^m)^n(b^n)^m = e^n e^m = ee = e$. So, $|ab|$ divides mn. Since mn is odd, so is $|ab|$.

69. In Z_6, $H = \{0, 1, 3, 5\}$ is not closed.

70. $1 \in H$. Let $a, b \in H$ and let m and n be integers such that $a^m = \pm 1$ and $b^m = \pm 1$. Then, $(ab)^{mn} = (a^m)^n(b^n)^{-m} = (\pm 1)^n(\pm 1)^{-m} = \pm 1$.

71. a. Let xh_1x^{-1} and xh_2x^{-1} belong to xHx^{-1}. Then $(xh_1x^{-1})(xh_2x^{-1})^{-1} = xh_1h_2^{-1}x^{-1} \in xHx^{-1}$ also.

b. Let $\langle h \rangle = H$. Then $\langle xhx^{-1} \rangle = xHx^{-1}$.

c. $(xh_1x^{-1})(xh_2x^{-1}) = xh_1h_2x^{-1} = xh_2h_1x^{-1} = (xh_2x^{-1})(xh_1x^{-1})$.

72. The identity is in H. Let $A, B \in H$. Then, det $A = m$ and det $B = n$ are positive rationals. Thus, det $(AB^{-1}) = (\text{det } A)(\text{det } B)^{-1} = m/n$, which is a positive rational.

73. Let a/b and c/d belong to the set. By observation, ac/bd and b/a have odd numerators and denominators. If ac/bd reduces to lowest terms to x/y, then x divides ac and y divides bd. So they are odd.

74. The identity has determinant $2^0 = 1$. Let $\det A = 2^m$ and $\det B = 2^n$. Then, $\det(AB) = 2^{m+n}$ and $\det A^{-1} = 2^{-m}$.

75. If 2^a and $2^b \in K$, then $2^a(2^b)^{-1} = 2^{a-b} \in K$, since $a - b \in H$.

76. The function $i(x) = 1$ where $x \neq 0$ is in H. Let $f, g \in H$. Then, $(f \cdot g)(2) = f(2) \cdot g(2) = 1 \cdot 1 = 1$. Also, $1 = i(2) = (f \cdot f^{-1})(2) = f(2) \cdot f^{-1}(2) = f^{-1}(2)$ so $f^{-1}(2) = 1$ and $f^{-1}(x)$ is in H. The 2 can be replaced by any nonzero number.

77. $\begin{bmatrix} 2 & 0 \\ 0 & 2 \end{bmatrix}^{-1} = \begin{bmatrix} \frac{1}{2} & 0 \\ 0 & \frac{1}{2} \end{bmatrix}$ is not in H.

78. D_n when n is odd; D_{n-1} when n is even.

79. If $a + bi$ and $c + di \in H$, then $(a+bi)(c+di)^{-1} = \frac{a+bi}{c+di}\frac{c-di}{c-di} = \frac{(ac+bd)+(bc-ad)i}{c^2+d^2} = (ac+bd) + (bc-ad)i$. Moreover, $(ac+bd)^2 + (bc-ad)^2 = a^2c^2 + 2acbd + b^2d^2 + b^2c^2 - 2bcad + a^2d^2$. Simplifying we obtain $(a^2+b^2)c^2 + (a^2+b^2)d^2 = (a^2+b^2)(c^2+d^2) = 1 \cdot 1 = 1$. So, H is a subgroup. H is the unit circle in the complex plane.

80. $a \in \langle a, b \rangle$. Let $a^i b^j$ and $a^s b^t \in \langle a, b \rangle$. Then, $(a^i b^j)(a^s b^t)^{-1} = a^{i-s} b^{j-t} \in \langle a, b \rangle$. $|\langle a, b \rangle| \leq |a||b|$.

81. $\{1, 2n-1, 2n+1, 4n-1\}$. This group is not cyclic.

82. Let F_1 and F_2 be distinct reflections in D_3 and $H = \{R_0, F_1\}$ and $K = \{R_0, F_2\}$. Then, $HK = \{R_0, F_2, F_1, F_1F_2\}$ but F_2F_1 is not in HK.

83. In D_{10} let a be any reflection and $b = R_{36}$.

84. First, observe that $p^n - 1$ is not divisible by p. For, if so, then $p^n - (p^n - 1) = 1$ would be divisible by p. So, $p^n - 1 \in U(p^n)$. Moreover, $(p^n-1)^2 = p^{2n} - 2p^n + 1 = 1$ mod p^n, so $p^n - 1$ has order 2. Now suppose that $x \in U(p^n)$ and $x^2 = 1$ mod p^n. We will prove that $x = 1$ or $x = p^n - 1$. From $x^2 = 1$ mod p^n we have that $x^2 - 1 = (x-1)(x+1)$ is divisible by p^n. We claim that p^n divides $x - 1$ or p^n divides $x + 1$. If this is not the case, then p divides both $x - 1$ and $x + 1$, and therefore p divides $(x+1) - (x-1) = 2$, which is false. Since $1 \leq x \leq p^n - 1$ we know that p^n divides $x - 1$ only when $x = 1$. Thus, if $x \neq 1$, we have $x + 1 \leq p^n$ and p^n divides $x + 1$. So, $x + 1 = p^n$.

85. First observe that $2^n - 1$ and $2^{n-2} \pm 1$ are in $U(2^n)$ and satisfy $x^2 = 1$. Now suppose that $x \in U(2^n)$, $x \neq 1$, and $x^2 = 1$ mod 2^n. From $x^2 = 1$ mod 2^n we have that $x^2 - 1 = (x-1)(x+1)$ is divisible by 2^n. Since $x - 1$ and $x + 1$ are even and $n \geq 3$, we know that at least one of $x - 1$ and $x + 1$ is divisible by 4. Moreover, it cannot be the case that both $x - 1$ and $x + 1$ are divisible by 4 for then so would $(x+1) - (x-1) = 2$. If $x - 1$ is not divisible by 4, then $x + 1$ is divisible by 2^{n-1}. Thus $x + 1 = k2^{n-1}$ for some integer k and $k2^{n-1} = x + 1 \leq 2^n$. So, $k = 1$ or $k = 2$. For $k = 1$, we have $x = 2^{n-1} - 1$. For $k = 2$, we have $x = 2^n - 1$. If $x + 1$ is not divisible by 4, then $x - 1$ is divisible by 2^{n-1}. Thus $x - 1 = k2^{n-1}$ for some integer k and $k2^{n-1} = x - 1 < 2^n$. So, $k = 1$ and $x = 2^{n-1} + 1$.

86. **a.** $U(5)$ or in $\mathbf{C}^*$ the subgroup $\{1, -1, i, -i\}$; $\mathbf{R}^*$

b. $GF(2, Z_3)$; $GF(2, Q)$

c. $U(8)$ or $U(12)$

d. Z_6

87. Since $ee = e$ is in $HZ(G)$ it is non-empty. Let h_1z_1 and h_2z_2 belong to $HZ(G)$. Then $h_1z_1(h_2z_2)^{-1} = h_1z_1z_2^{-1}h_2^{-1} = h_1h_2^{-1}z_1z_2^{-1} \in HZ(G)$.

88. Observe that if $a/b \in H$ and $c/d \in K$ where a, b, c, d are nonzero intergers, then $a = b(a/b) \in H,\ c = d(c/d) \in K$ and $ac \in H \cap K$.

89. First note that if $m/n \neq 0$ is an element of H, then $n(m/n) = m$ and $-m$ are also in H. By the Well Ordering Principle, H has a least positive integer t. Since t is not in $K = \{2h \mid h \in H\}$, K is a nontrivial proper subgroup of H (see Example 5). Alternatively, one can use Exercise 88.

90. Suppose that G is a group of order $n > 2$ and H is a subgroup of G with $|H| = n - 1$. Let a be in G but a not in H and let b be in H and $b \neq e$. Then, $ab \neq a$ and ab is not in H.

CHAPTER 4

Cyclic Groups

1. For Z_6, generators are 1 and 5; for Z_8 generators are 1, 3, 5, and 7; for Z_{20} generators are 1, 3, 7, 9, 11, 13, 17, and 19.

2. For $\langle a \rangle$, generators are a and a^5; for $\langle b \rangle$, generators are b, b^3, b^5, and b^7; for $\langle c \rangle$, generators are c, c^3, c^7, c^9, c^{11}, c^{13}, c^{17}, c^{19}.

3. $\langle 20 \rangle = \{20, 10, 0\}$; $\langle 10 \rangle = \{10, 20, 0\}$
   $\langle a^{20} \rangle = \{a^{20}, a^{10}, a^0\}$; $\langle a^{10} \rangle = \{a^{10}, a^{20}, a^0\}$

4. $\langle 3 \rangle = \{3, 6, 9, 12, 15, 0\}$; $\langle 15 \rangle = \{15, 12, 9, 6, 3, 0\}$; $\langle a^3 \rangle = \{a^3, a^6, a^9, a^{12}, a^{15}, a^0\}$; $\langle a^{15} \rangle = \{a^{15}, a^{12}, a^9, a^6, a^3, a^0\}$.

5. $\langle 3 \rangle = \{3, 9, 7, 1\}$
   $\langle 7 \rangle = \{7, 9, 3, 1\}$

6. In any group, $\langle a \rangle = \langle a^{-1} \rangle$. See Exercise 11.

7. This follows directly from Corollary 2 of Theorem 4.1.

8. (a) All have order 5. (b) Both have order 3. (c) All have order 15.

9. Six subgroups; generators are the divisors of 20.
   Six subgroups; generators are a^k, where k is a divisor of 20.

10. $3 \cdot 1, 3 \cdot 3, 3 \cdot 5, 3 \cdot 7; a^3, (a^3)^3, (a^3)^5, (a^3)^7$.

11. By definition, $a^{-1} \in \langle a \rangle$. So, $\langle a^{-1} \rangle \subseteq \langle a \rangle$. By definition, $a = (a^{-1})^{-1} \in \langle a^{-1} \rangle$. So, $\langle a \rangle \subseteq \langle a^{-1} \rangle$.

12. $\langle 3 \rangle, \langle -3 \rangle; a^3,\ a^{-3}$.

13. Observe that $\langle 10 \rangle = \{0, \pm 10, \pm 20, \ldots\}$ and $\langle 12 \rangle = \{0, \pm 12, \pm 24, \ldots\}$.
    Since the intersection of two subgroups is a subgroup, according to the proof of Theorem 4.3, we can find a generator of the intersection by taking the smallest positive integer that is in the intersection. So, $\langle 10 \rangle \cap \langle 12 \rangle = \langle 60 \rangle$. For m and n we have $\langle m \rangle = \{0, \pm m, \pm 2m, \ldots\}$ and $\langle n \rangle = \{0, \pm n, \pm 2n, \ldots\}$. Then the smallest positive integer in the intersection is $\text{lcm}(m, n)$.

    For the case $\langle a^m \rangle \cap \langle a^n \rangle$, let $k = \text{lcm}(m, n)$. Write $k = ms$ and $k = nt$. Then $a^k = (a^m)^s \in \langle a^m \rangle$ and $a^k = (a^n)^t \in \langle a^n \rangle$. So, $\langle a^k \rangle \subseteq \langle a^m \rangle \cap \langle a^n \rangle$. Now let a^r be any element in $\langle a^m \rangle \cap \langle a^n \rangle$. Then r is a multiple of both m and n. It follows that r is a multiple of k (see Exercise 12 of Chapter 0). So, $a^r \in \langle a^k \rangle$.

14. 49. First note that the group is not infinite since an infinite cyclic group has infinitely many subgroups. Let $|G| = n$. Then 7 and $n/7$ are both divisors of n. If $n/7 \neq 7$, then G has at least 4 divisors. So, $n/7 = 7$. When 7 is replaced by p , $|G| = p^2$.

15. $|g|$ divides 12 is equivalent to $g^{12} = e$. So, if $a^{12} = e$ and $b^{12} = e$, then $(ab^{-1})^{12} = a^{12}(b^{12})^{-1} = ee^{-1} = e$. The same argument works when 12 is replaced by any integer (see Exercise 51 of Chapter 3).

16. **a.** $|a| = |a^2|$ if and only if $|a|$ is odd or infinite. To see this note, that if $|a| = \infty$, then $|a^2|$ cannot be finite, and if $|a| = n$, by Theorem 4.2, we have $n = |a^2| = n/\gcd(n, 2)$ and therefore $\gcd(n, 2) = 1$. **b.** $|a^2| = |a^{12}|$ if and only if $|a| = \infty$ or $|a|$ is finite and $\gcd(|a|, 2) = \gcd(|a|, 12)$. **c.** Both i and j are 0 or both are not 0. **d.** $i = \pm j$.

17. By Theorem 4.2, we have $|\langle a^6 \rangle| = n/\gcd(n, 6)$. Since n is odd and $\langle a^6 \rangle$ is a proper subgroup, we have $\gcd(n, 6) = 3$. So, $|\langle a^6 \rangle| = n/3$.

18. By Theorem 4.2, we have $\gcd(n, 10) = \gcd(n, 5)$. If n is even and $\gcd(n, 5) = 5$, then $\gcd(n, 10) = 10$, which contradicts that $\gcd(n, 10) = \gcd(n, 5)$.

19. If $|a^2| = 3$, $|a|$ is 3 or 6. If $|a^2| = 4$, $|a| = 8$.

20. For D_{p^n} there are p^n cyclic subgroups of order 2. Since the rotations form a cyclic subgroup of order p^n there is exactly one subgroup for each of the orders $p^0, p^1, p^2, \ldots, p^n$ and no others. So, the total for D_{p^n} is $p^n + n + 1$. For D_{pq} there are pq cyclic subgroups of order 2. Since the rotations form a cyclic subgroup of order pq, there is exactly one cyclic subgroup for each of the orders pq, p, q and 1. So, the total for D_{pq} is $pq + 4$.

21. For every a and b we have $ab = (ab)^{-1} = b^{-1}a^{-1} = ba$.
Alternate solution. Let a and b belong to G. Observe that $aabb = a^2b^2 = ee = e = (ab)^2 = abab$. By cancellation we have $ab = ba$.

22. $\phi(81) = 27 \cdot 2 = 54$; $\phi(60) = \phi(4)\phi(3)\phi(5) = 2 \cdot 2 \cdot 4 = 16$; $\phi(105) = \phi(3) \cdot \phi(5) \cdot \phi(7) = 2 \cdot 4 \cdot 6 = 48$.

23. Let $|a| = m$, $|b| = n$, $|ab| = k$ and $\gcd(m, n) = d$. Then $\operatorname{lcm}(m, n) = mn/d$ and $(ab)^{mn/d} = (a^m)^{n/d}(b^{n/d})^m = ee = e$ so k divides $\operatorname{lcm}(m, n)$. So, if $d > 1$, then $k < mn$. If $d = 1$, then $\langle a \rangle \cap \langle b \rangle = \{e\}$ because $|\langle a \rangle \cap \langle b \rangle|$ divides both $|\langle a \rangle|$ and $|\langle b \rangle|$. We also have $e = (ab)^k = a^k b^k$ and therefore $a^k = b^{-k} \in \langle a \rangle \cap \langle b \rangle = \{e\}$. This means that both m and n and therefore mn are divisors of k.

24. First suppose G is infinite. Let $x \in G, x \neq e$. Then, $G = \langle x \rangle$ and $\langle x^2 \rangle$ contradicts the hypothesis. Next, assume G is finite and $e \neq x \in G$. Then, $\langle x \rangle = G$ (otherwise $\langle x \rangle$ is nontrivial and proper). By Theorem 4.3, G has a subgroup for each divisor of $|G|$ and since the only subgroups of G have orders $|G|$ and 1 we have that only divisors of $|G|$ are $|G|$ and 1. So, $|G|$ is prime.

25. Exercise 31 in Chapter 3 tells us that H is a subgroup of the cyclic group of n rotations in D_n. So, by Theorem 4.3, H is cyclic.

26. Z_{3n}; D_{3n}. These generalize to the p odd case.

27. 1 (the identity). To see this, note that we can let the group be $\langle a \rangle$ where $|a|$ is infinite. If some element a^i has finite order n then $(a^i)^n = e$. But then $a^{in} = e$, which implies that a has finite order. This contradicts our assumption.

28. By Corollary 2 of Theorem 4.1, a nonidentity element of G must have order 5, 7 or 35. We may assume that G has no element of order 35. Since 34 is not a multiple of $\phi(5) = 4$, not all of the nonidentity elements can have order 5. Similarly, not all of them can have order 7. So, G has elements of orders both 5 and 7. Say, $|a| = 5$ and $|b| = 7$. Then, since $(ab)^5 = b^5 \neq e$ and $(ab)^7 = a^7 = a^2 \neq e$, we must have $|ab| = 35$, a contradiction.

29. **a.** $|a|$ divides 12. **b.** $|a|$ divides m. **c.** By Theorem 4.3, $|a| = 1, 2, 3, 4, 6, 8, 12$, or 24. If $|a| = 2$, then $a^8 = (a^2)^4 = e^4 = e$. A similar argument eliminates all other possibilities except 24.

30. Let $G = \{e, a, b\}$. Cancellation shows ab must be e. Thus, $G = \{e, a, a^{-1}\} = \langle a \rangle$.

31. Yes, by Theorem 4.3. The subgroups of Z are of the form $\langle n \rangle = \{0, \pm n, \pm 2n, \pm 3n, \ldots\}$, for $n = 0, 1, 2, 3, \ldots$. The subgroups of $\langle a \rangle$ are of the form $\langle a^n \rangle$ for $n = 0, 1, 2, 3, \ldots$

32. Certainly, $a \in C(a)$. Thus, $\langle a \rangle \subseteq C(a)$.

33. D_n has n reflections, each of which has order 2. D_n also has n rotations that form a cyclic group of order n. So, according to Theorem 4.4, there are $\phi(d)$ rotations of order d in D_n. If n is odd, there are no rotations of order 2. If n is even, there is $\phi(2) = 1$ rotation of order 2. (Namely, R_{180}.) So, when n is odd, D_n has n elements of order 2; when n is even, D_n has $n + 1$ elements of order 2.

34. 1 and -1 are the only generators of Z. Suppose that a^k generates $\langle a \rangle$. Then there is an integer t so that $(a^k)^t = a$. By Theorem 4.1, we conclude that $kt = 1$. So, $k = \pm 1$.

35. See Example 16 of Chapter 2.

36. Let a and b to H. Then, $a \in H_i$ and $b \in H_j$ for some i and j and we may assume that $i < j$. Then, $ab^{-1} \in H_j \subset H$.

37. 1000000, 3000000, 5000000, 7000000. By Theorem 4.3, $\langle 1000000 \rangle$ is the unique subgroup of order 8, and only those on the list are generators; $a^{1000000}, a^{3000000}, a^{5000000}, a^{7000000}$. By Theorem 4.3, $\langle a^{1000000} \rangle$ is the unique subgroup of order 8, and only those on the list are generators.

38. Let H be the unique maximal subgroup of G and suppose that G is not cyclic. Let a be any element in G not in H. Then $H_1 = \langle a \rangle$ is not G. Since H is the unique maximal subgroup of G, H_1 cannot be maximal and is therefore properly contained some proper subgroup H_2 of G. Because $a \in H_2$, H_2 cannot be maximal. Since G is finite, we can repeat this argument to reach a contradiction. Thus, $G = \langle a \rangle$. Now suppose that $|a| = n$ and n is divisible by two distinct primes p and q. Then, $\langle a^{m/p} \rangle$ and $\langle a^{m/q} \rangle$ are distinct maximal subgroups of G, which is a contradiction.

39. Let $G = \{a_1, a_2, \ldots, a_k\}$. Now let $|a_i| = n_i$ and $n = n_1 n_2 \ldots n_k$. Then $a_i^n = e$ for all i since n is a multiple of n_i.

40.

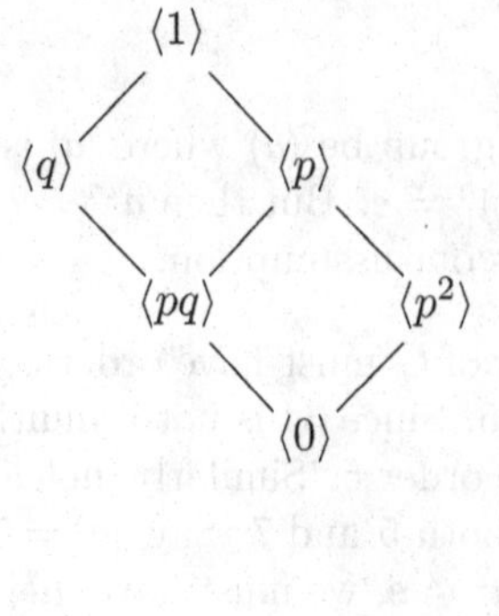

41. The lattice is a vertical line with successive terms from top to bottom $\langle p^0\rangle, \langle p^1\rangle, \langle p^2\rangle, \ldots, \langle p^{n-1}\rangle, \langle 0\rangle$.

42. First suppose that G is the union of proper subgroups. If G were cyclic, say $G = \langle a\rangle$, and G was the union of proper subgroups $H_1, H_2, \ldots, H_n$, then a must be in one of H_i, since the union contains every element. But if a belongs to H_i, then $G = \langle a\rangle$ is a subgroup of the proper subgroup H_i. This is a contradiction. Now suppose that G is not cyclic. Then, for every $g \in G$, $\langle g\rangle$ is a proper subgroup and $G = \cup_{g\in G}\langle g\rangle$.

43. Suppose that Q^+ is cyclic. Because $\langle a/b\rangle = \langle b/a\rangle$ we may assume that $a/b > 1$. Let p be any prime that does not divide a. Then there is a positive integer such that $(a/b)^n = p$. Thus $a^n = pb^n$. But this contradicts Theorem 0.3.
Alternate solution. Suppose that r is a generator of Q^+. Since $\langle r\rangle = \langle r^{-1}\rangle$, we may assume that $r > 1$. Then there are positive integers m and n such that $r^m = 2$ and $r^n = 3$. Then $r^{mn} = (r^m)^n = 2^n$ and $r^{mn} = (r^n)^m = 3^m$. This implies that $2^n = 3^m$. But 2^n is even and 3^m is odd. This proves the group of nonzero rationals under multiplication is not cyclic, for otherwise, its subgroups would be cyclic.

44.

	4	8	12	16
4	16	12	8	4
8	12	4	16	8
12	8	16	4	12
16	4	8	12	16

The identity is 16. The group is generated by 8 and by 12.

45. For 7, use Z_{2^6}. For n, use $Z_{2^{n-1}}$.

46. $|ab|$ could be any divisor of $\text{lcm}(|a|, |b|)$.

47. Suppose that $|ab| = n$. Then $(ab)^n = e$ implies that $b^n = a^{-n} \in \langle a\rangle$, which is finite. Thus, $|b|$ is finite.

48. See Example 5 in Chapter 3 for showing G is a subgroup. For the general case, observe by Theorem 4.2 for any $a \in G$ we have $\langle a^k\rangle = \langle a^{\gcd(n,k)}\rangle$ so that $G^k = G^{\gcd(n,k)}$.

49. Since $\gcd(100, 98) = 2$ and $\gcd(100, 70) = 10$ we have $|a^{98}| = |a^2| = 50$ and $|a^{70}| = |a^{10}| = 10$.

50. Since FF' is a rotation other than the identity and the rotations of D_{21} form a cyclic subgroup of order 21, we know by Theorem 4.3 that $|FF'|$ is a divisor of 21. Moreover, FF' cannot be the identity for then $FF' = FF$, which implies that $F' = F$. So, $|FF'| = 3, 7$ or 21.

51. Because H is cyclic, we know that $|a^6|$ divides 10. So, $a^{60} = e$. Thus $|a|$ can be any divisor of 60.

52. Using the corollary to Theorem 4.4 we get 21600

53. The argument given in the proof of the corollary to Theorem 4.4 shows that in an infinite group, the number of elements of finite order n is a multiple of $\phi(n)$ or there is an infinite number of elements of order n.

54. By Corollary 1 of Theorem 4.2, $|a| = m$ divides n. Say $n = mq$. Then, $a^n = (a^m)^q = e^m = e$.

55. It follows from Example 16 in Chapter 2 and Example 15 in Chapter 0 that the group $H = \langle \cos(360°/n) + i\sin(360°/n) \rangle$ is a cyclic group of order n and every member of this group satisfies $x^n - 1 = 0$. Moreover, since every element of order n satisfies $x^n - 1 = 0$ and there can be at most n such elements, all complex numbers of order n are in H. Thus, by Theorem 4.4, C^* has exactly $\phi(n)$ elements of order n.

56. Clearly 0 is in H. If m and n are in H, then m has the forms $8m_1$ and $10m_2$ and n has the forms $8n_1$ and $10n_2$. Then, $m - n$ has the forms $8m_1 - 8n_1$ and $10m_1 - 10n_1$. So, $m - n$ is in H and H is a subgroup of Z. If the condition is changed to "divisible by 8 or 10," H is not a subgroup since 8 and 10 would belong to H but $10 - 8 = 2$ would not.

57. Let $x \in Z(G)$ and $|x| = p$ where p is prime. Say $y \in G$ with $|y| = q$ where q is prime. Then $(xy)^{pq} = e$ and therefore $|xy| = 1, p$ or q. If $|xy| = 1$, then $x = y^{-1}$ and therefore $p = q$. If $|xy| = p$, then $e = (xy)^p = y^p$ and q divides p. Thus, $q = p$. A similar argument applies if $|xy| = q$.

58. If an infinite group had only a finite number of subgroups, then it would have only a finite number of cyclic subgroups. If each of these cyclic subgroups is finite then the group would be finite since every element is in the cyclic subgroup generated by itself. So the group contains at least one infinite cyclic subgroup, call it $\langle a \rangle$. Then the subgroups $\langle a \rangle, \langle a^2 \rangle, \langle a^3 \rangle, \ldots$ are distinct subgroups since the least positive power of a in $\langle a^i \rangle$ is a^i and a^i is not the least positive power of a in any subgroup $\langle a^j \rangle$ for $j \neq i$.

59. An infinite cyclic group does not have an element of prime order. A finite cyclic group can have only one subgroup for each divisor of its order. A subgroup of order p has exactly $p - 1$ elements of order p. Another element of order p would give another subgroup of order p.

60. 2; 4; a^3, a^5, a^7.

61. $1 \cdot 4$, $3 \cdot 4$, $7 \cdot 4$, $9 \cdot 4$; $a^4, (a^4)^3, (a^4)^7, (a^4)^9$.

62. In a group, the number of elements order d is divisible by $\phi(d)$ or there are infinitely many elements of order d.

63. D_{33} has 33 reflections, each of which has order 2 and 33 rotations that form a cyclic group. So, according to Theorem 4.4, for each divisor d of 33 there are $\phi(d)$ rotations of order d in D_n. This gives one element of order 1; $\phi(3) = 2$ elements of order 3; $\phi(11) = 10$ elements of order 11; and $\phi(33) = 20$ elements of order 33.

64. Since $U(25) = 20$, by Corollary 1 of Theorem 4.2, we know that $|2|$ must divide 20. So, $|2| = 1, 2, 4, 5, 10$, or 20. But $2^{10} \neq 1$ implies that $|2| \neq 1, 2, 5$ or 10 and $2^4 \neq 1$ implies that $|2| \neq 4$.

65. Let $|\langle a \rangle| = 4$ and $|\langle b \rangle| = 5$. Since $(ab)^{20} = (a^4)^5(b^5)^4 = e \cdot e = e$, we know that $|ab|$ divides 20. Noting that $(ab)^4 = b^4 \neq e$ we know that $|ab| \neq 1, 2$ or 4. Likewise, $(ab)^{10} = a^2 \neq e$ implies that $|ab| \neq 5$ or 10. So, $|ab| = 20$. Then, by Theorem 4.3, $\langle ab \rangle$ has subgroups of orders 1, 2, 4, 5, 10 and 20. In general, if an Abelian group contains cyclic subgroups of order m and n where m and n are relatively prime, then it contains subgroups of order d for each divisor d of mn.

66. 1, 2, 3, 12. In general, if an Abelian group contains cyclic subgroups of order m and n, then it contains subgroups of order d for each divisor d of the least common multiple of m and n.

67. Say a and b are distinct elements of order 2. If a and b commute, then ab is a third element of order 2. If a and b do not commute, then aba is a third element of order 2.

68. $\phi(42) = 12$

69. By Exercise 38 of Chapter 3, $\langle a\rangle \cap \langle b\rangle$ is a subgroup. Also, $\langle a\rangle \cap \langle b\rangle \subseteq \langle a\rangle$ and $\langle a\rangle \cap \langle b\rangle \subseteq \langle b\rangle$. So, by Theorem 4.3, $|\langle a\rangle \cap \langle b\rangle|$ is a common divisor of 10 and 21. Thus, $|\langle a\rangle \cap \langle b\rangle| = 1$ and therefore $\langle a\rangle \cap \langle b\rangle = \{e\}$.

70. Mimic Exercise 69.

71. $|\langle a\rangle \cap \langle b\rangle|$ must divide both 24 and 10. So, $\langle a\rangle \cap \langle b\rangle| = 1$ or 2.

72. By Exercise 38 of Chapter 3, $\langle a\rangle \cap \langle b\rangle$ is a subgroup $\langle a\rangle$ and $\langle b\rangle$. So, $|\langle a\rangle \cap \langle b\rangle|$ divides 12 and 22. It follows that $|\langle a\rangle \cap \langle b\rangle| = 1$ or 2, and since $\langle a\rangle \cap \langle b\rangle \neq \{e\}$, we have that $|\langle a\rangle \cap \langle b\rangle| = 2$. Because $\langle a^6\rangle$ is the only subgroup of $\langle a\rangle$ of order 2 and $\langle b^{11}\rangle$ is the only subgroup of $\langle b\rangle$ of order 2, we have $\langle a\rangle \cap \langle b\rangle = \langle a^6\rangle = \langle b^{11}\rangle$, and therefore $a^6 = b^{11}$.

73. Suppose that G has 14 elements of order 3. Let $a \in G$ and $a \neq e$. Let $b \in G$ and $b \notin \langle a\rangle$. Then, by cancellation, $H = \{a^i b^j \mid i, j \text{ are } 0, 1, 2\}$ has exactly nine elements and is closed and therefore is a subgroup of G. Let $c \in G$ and $c \notin H$. Then, by cancellation, the nine expressions of the form $a^i b^j c$ where i, j are $0, 1, 2$ are distinct and have no overlap with the nine elements of H. But that gives 18 elements in G.

74. First note that if k is a generator then so is $-k$. Thus, it suffices to show that $k \neq -k$. But $k = -k$ implies that $2k = 0$ so that $n = |k| = 1$ or 2.

75. Observe that $|a^5| = 12$ implies that $e = (a^5)^{12} = a^{60}$, so $|a|$ divides 60. Since $\langle a^5\rangle \subseteq \langle a\rangle$, we know that $|\langle a\rangle|$ is divisible by 12. So, $|\langle a\rangle| = 12$ or 60.

If $|a^4| = 12$, then $|a|$ divides 48. Since $\langle a^4\rangle \subseteq \langle a\rangle$, we know that $|\langle a\rangle|$ is divisible by 12. So, $|\langle a\rangle| = 12, 24$, or 48. But $|a| = 12$ implies $|a^4| = 3$ and $|a| = 24$ implies $|a^4| = 6$. So, $|a| = 48$.

76. By Theorem 4.3, it suffices to find necessary and sufficient conditions so that $|x^r|$ divides $|x^s|$. By Theorem 4.2, we obtain $\gcd(n, s)$ divides $\gcd(n, r)$.

77. gcd(48,21) = 3; gcd(48,14) = 2; gcd(48,18) = 6.

78. $\{R_0, R_{90}, R_{180}, R_{270}\}$; $\{R_0, R_{180}, F, R_{180}F\}$ where F is any reflection.

79. $e \in H$. Let $a, b \in H$, $|a| = m, |b| = n$. Then $(ab^{-1})^{mn} = (a^m)^n (b^n)^m = e$. So, $|ab^{-1}|$ divides mn, which is odd. So, $|ab^{-1}|$ is odd.

80. In Z_6 $H = \{0, 1, 3, 5\}$ but $3 + 5 = 2$, which has order 3.

81. Since $ee = e$ is in $HZ(G)$ it is non-empty. Let $h_1 z_1$ and $h_2 z_2$ belong to $HZ(G)$. Then $h_1 z_1 (h_2 z_2)^{-1} = h_1 z_1 z_2^{-1} h_2^{-1} = h_1 h_2^{-1} z_1 z_2^{-1} \in HZ(G)$.

82. Note that n and $n^2 - 2$ are distinct elements of $U(n^2 - 1)$ of order 2 and appeal to Theorem 4.4.

83. Observe that $n^2 - 2 = -1$ and n are in $U(n^2 - 1)$ and have order 2. Thus $\{\pm 1, \pm n\}$ is closed and therefore is a subgroup.

84. By Theorem 4.4 a cyclic group of even order must have a unique element or order 2 whereas $U(2^n)$ $(n \geq 3)$ has $2^{n-1} \pm 1$ and $-1 = 2^n - 1$.

85. Note that among the integers from 1 to p^n the p^{n-1} integers $p, 2p, \ldots, p^{n-1}p$ are exactly the ones not relatively prime to p.

86. Observe that $|U(n)| = \phi(n)$. So, $|U(|U(n)|) = |U(\phi(n))| = \phi(\phi(n))$.

CHAPTER 5

Permutation Groups

1. a. $\alpha^{-1} = \begin{bmatrix} 1 & 2 & 3 & 4 & 5 & 6 \\ 2 & 1 & 3 & 5 & 4 & 6 \end{bmatrix}$

   b. $\beta\alpha = \begin{bmatrix} 1 & 2 & 3 & 4 & 5 & 6 \\ 1 & 6 & 2 & 3 & 4 & 5 \end{bmatrix}$

   c. $\alpha\beta = \begin{bmatrix} 1 & 2 & 3 & 4 & 5 & 6 \\ 6 & 2 & 1 & 5 & 3 & 4 \end{bmatrix}$

2. $\alpha = (12345)(678) = (15)(14)(13)(12)(68)(67); \beta = (23847)(56) = (27)(24)(28)(23)(56); \alpha\beta = (12485736) = (16)(13)(17)(15)(18)(14)(12)$.

3. a. (15)(234) b. (124)(35)(6) c. (1423)

4. 2; 3; 5; k.

5. a. lcm(3,3) = 3.
   b. lcm(3,4) = 12.
   c. lcm(3,2) = 6.
   d. lcm(3,6) = 6.
   e. $|(1235)(24567)| = |(124)(3567)| = \text{lcm}(3,4) = 12$.
   f. $|(345)(245)| = |(25)(34)| = \text{lcm}(2,2) = 2$.

6. 6; 12

7. By Theorem 5.3 the orders are lcm(4,6) = 12 and lcm(6,8,10) = 120.

8. (135) =(15)(13) even; (1356) = (16)(15)(13) odd; (13567) = (17)(16)(15)(13) even; (12)(134)(152) = (12)(14)(13)(12)(15) odd; (1243)(3521) = (13)(14)(12)(31)(32)(35) even.

9. $((14562)(2345)(136)(235))^{10} = ((153)(46))^{10} = (135)^{10}(46)^{10} = (153)$.

10. (13)(1245)(13) = (3245); (24)(13456)(24) = (13256). In general, for any cycle α we have $(ij)\alpha(ij)$ is the same as α with i replaced by j. In both cases, the element of the 2-cycle that appears in all three cycles is replaced by the other element of the 2-cycle.

11. An n-cycle is even when n is odd, since we can write it as a product of $n-1$ 2-cycles by successively pairing up the first element of the cycle with each of the other cycle elements starting from the last element of the cycle and working toward the front. The same process shows that when n is odd we get an even permutation.

12. If $\alpha_1, \alpha_2, \ldots, \alpha_n$ are 2-cycles and $\alpha = \alpha_1 \cdots \alpha_n$, then $\alpha^{-1} = \alpha_n\alpha_{n-1} \cdots \alpha_2\alpha_1$.

13. To prove that α is $1-1$, assume $\alpha(x_1) = \alpha(x_2)$. Then $x_1 = \alpha(\alpha(x_1)) = \alpha(\alpha(x_2)) = x_2$. To prove that α is onto, note that for any s in S, we have $\alpha(\alpha(s)) = s$.

14. $(n-3)!$ in S_n; $(n-3)!/2$ in A_n.

15. Suppose that α can be written as a product on m 2-cycles and β can be written as product of n 2-cycles. Then juxtaposing these 2-cycles we can write $\alpha\beta$ as a product of $m+n$ 2-cycles. Now observe that $m+n$ is even if and only if m and n are even or both odd.

16. $(+1) \cdot (+1) = (+1)$ $\quad$ $(-1) \cdot (-1) = +1$
even · *even* = *even* $\quad$ *odd* · *odd* = *even*

$(+1) \cdot (-1) = (-1)$ $\quad$ $(-1) \cdot (+1) = (-1)$
even · *odd* = *odd* $\quad$ *odd* · *even* = *odd*

17. n is odd

18. even; odd.

19. If α is the product of m 2-cycles and β is the product of n 2-cycles, then $\alpha^{-5}\beta\alpha^3$ is the product of $8m+n$ and $8m+n$ is odd if and only if n is odd.

20. Say β can be written with m 2-cycles and α with n 2-cycles. Then, $\beta^{-1}\alpha\beta$ can be written with $2m+n$ 2-cycles.

21. even

22. 10; 12

23. We find the orders by looking at the possible products of disjoint cycle structures arranged by longest lengths left to right and denote an n-cycle by $(\underline{n})$.
$(\underline{6})$ has order 6 and is odd;
$(\underline{5})(\underline{1})$ has order 5 and is even;
$(\underline{4})(\underline{2})$ has order 4 and is even;
$(\underline{4})(\underline{1})(\underline{1})$ has order 4 and is odd;
$(\underline{3})(\underline{3})$ has order 3 and is even;
$(\underline{3})(\underline{2})(\underline{1})$ has order 6 and is odd;
$(\underline{3})(\underline{1})(\underline{1})(\underline{1})$ has order 3 and is even;
$(\underline{2})(\underline{2})(\underline{2})$ has order 2 and is odd;
$(\underline{2})(\underline{2})(\underline{1})(\underline{1})$ has order 2 and is even;
$(\underline{2})(\underline{1})(\underline{1})(\underline{1})(\underline{1})$ has order 2 and is odd.
So, for S_6, the possible orders are 1, 2, 3, 4, 5, 6; for A_6 the possible orders are 1, 2, 3, 4, 5. We see from the cycle structure of S_7 shown in Example 4 that in A_7 the possible orders are 1, 2, 3, 4, 5, 6, 7.

24. (123)(45678; (12)(345)(6, 7, 8, 9, 10)(11, 12)

25. Since $|\beta| = 21$, we have $n = 16$.

26. $(a_n a_{n-1} \cdots a_2 a_1)$

27. If all members of H are even we are done. So, suppose that H has at least one odd permutation σ. For each odd permutation β in H observe that $\sigma\beta$ is even and, by cancellation, different βs give different $\sigma\beta$s. Thus, there are at least as many even permutations as there are odd ones. Conversely, for each even permutation β in H observe that $\sigma\beta$ is odd and, by cancellation, different βs give different $\sigma\beta$s. Thus, there are at least as many odd permutations as there are even ones.

28. By Exercise 27, either every element of H is even or half are even and half are odd. In the latter case, H would have even order.

29. The identity is even; the set is not closed.

30. If α can be written as the product of m 2-cycles and β can be written as a product n 2-cycles, then $\alpha^{-1}\beta^{-1}\alpha\beta$ can be written as the product of $n+m+n+m = 2(m+n)$ 2-cycles.

31. $(8 \cdot 7 \cdot 6 \cdot 5 \cdot 4 \cdot 3 \cdot 2 \cdot 1)/(2 \cdot 2 \cdot 2 \cdot 2 \cdot 4!)$

32. $(7 \cdot 6 \cdot 5 \cdot 4 \cdot 3)/5$

33. In A_6, elements of order 2 in disjoint cycle form must be the product of two 2-cycles. So the number of elements of order 2 is $6 \cdot 5 \cdot 4 \cdot 3/(2 \cdot 2 \cdot 2)$.

34. Suppose that $\alpha \in Z(A_n)$ and $\alpha \neq \epsilon$. Pick $i \in \{1, 2, \ldots, n\}$ such that $\alpha(i) = j \neq i$ and let $\alpha(j) = k$. Let $t \in \{1, 2, \ldots, n\}$ and $t \neq i, j$, or k. If $k = i$, then $(jti)\alpha = \alpha(jti)$. But the left side sends i to t, whereas the right side sends i to i, which is a contradiction. If $k \neq i$, then the left side of $(jti)\alpha = \alpha(jti)$ sends i to t, whereas the right side sends i to k, which is a contradiction.

35. Since $|x^5| = 5$, we know $|x| = 25$. One solution is $(1, 6, 7, 8, 9, 2, 10, 11, 12, 13, 3, 14, 15, 16, 17, 4, 18, 19, 20, 21, 5, 22, 23, 24, 25)$. The number of solutions is 20!

36. Since $\alpha^m = (1, 3, 5, 7, 9)^m(2, 4, 6)^m(8, 10)^m$ and the result is a 5-cycle, we deduce that $(2, 4, 6)^m = \epsilon$ and $(8, 10)^m = \epsilon$. So, 3 and 2 divide m. Since $(1, 3, 5, 7, 9)^m \neq \epsilon$ we know that 5 does not divide m. Thus, we can say that m is a multiple of 6 but not a multiple of 30.

37. An odd permutation of order 4 must be of the form $(a_1a_2a_3a_4)$. There are 6 choices for a_1, 5 for a_2, 4 for a_3, and 3 for a_4. This gives $6 \cdot 5 \cdot 4 \cdot 3$ choices. But since for each of these choices the cycles $(a_1a_2a_3a_4) = (a_2a_3a_4a_1) = (a_3a_4a_1a_2) = (a_4a_3a_2a_1)$ give the same group element, we must divide $6 \cdot 5 \cdot 4 \cdot 3$ by 4 to obtain 90. An even permutation of order 4 must be of the form $(a_1a_2a_3a_4)(a_5a_6)$. As before, there are 90 choices $(a_1a_2a_3a_4)$. Since $(a_5a_6) = (a_6a_5)$ there are 90 elements of the form $(a_1a_2a_3a_4)(a_5a_6)$. This gives 180 elements of order 4 in S_6.

    A permutation in S_6 of order 2 has three possible disjoint cycle forms: (a_1a_2), $(a_1a_2)(a_3a_4)$ and $(a_1a_2)(a_3a_4)(a_5a_6)$. For (a_1a_2) there are $6 \cdot 5/2$ =15 distinct elements; for $(a_1a_2)(a_3a_4)$ there are $6 \cdot 5 \cdot 4 \cdot 3$ choices for the four entries but we must divide by $2 \cdot 2 \cdot 2$ since $(a_1a_2) = (a_2a_1), (a_3a_4) = (a_4a_3)$ and $(a_1a_2)(a_3a_4) = (a_4a_3)(a_1a_2)$. This gives 45 distinct elements. For $(a_1a_2)(a_3a_4)(a_5a_6)$ there are 6! choices for the six entries but we must divide by $2 \cdot 2 \cdot 2 \cdot 3!$ since each of the three 2-cycles can be written 2 ways and the three 2-cycles can be permuted 3! ways. This gives 15 elements. So, the total number of elements of order 2 is 75.

38. Any product of 3-cycles is even, whereas (1234) is odd. In general, no odd permutation can be written as a product of 3-cycles.

39. Since $\beta^{28} = (\beta^4)^7 = \epsilon$, we know that $|\beta|$ divides 28. But $\beta^4 \neq \epsilon$ so $|\beta| \neq 1, 2$, or 4. If $|\beta| = 14$, then β written in disjoint cycle form would need at least one 7-cycle and one 2-cycle. But that requires at least 9 symbols and we have only 7. Likewise, $|\beta| = 28$ requires at least one 7-cycle and one 4-cycle. So, $|\beta| = 7$. Thus, $\beta = \beta^8 = (\beta^4)^2 = (2457136)$. In S_9, $\beta = (2457136)$ or $\beta = (2457136)(89)$.

40. Observe that $\beta = (123)(145) = (14523)$ so that $\beta^{99} = \beta^4 = \beta^{-1} = (13254)$.

41. Since $|(a_1a_2a_3a_4)(a_5a_6)| = 4$, such an x would have order 8. But the elements in S_{10} of order 8 are 8-cycles or the disjoint product of an 8-cycle and a 2-cycle. In both cases the square of such an element is the product of two 4-cycles.

42. If α and β are disjoint 2-cycles, then $|\alpha\beta| = \text{lcm}(2,2) = 2$. If α and β have exactly one symbol in common we can write $\alpha = (ab)$ and $\beta = (ac)$. Then $\alpha\beta = (ab)(ac) = (acb)$ and $|\alpha\beta| = 3$.

43. Let $\alpha, \beta \in \text{stab}(a)$. Then $(\alpha\beta)(a) = \alpha(\beta(a)) = \alpha(a) = a$. Also, $\alpha(a) = a$ implies $\alpha^{-1}(\alpha(a)) = \alpha^{-1}(a)$ or $a = \alpha^{-1}(a)$.

44. Let $\beta, \gamma \in H$. Then $(\beta\gamma)(1) = \beta(\gamma(1)) = \beta(1) = 1$; $(\beta\gamma)(3) = \beta(\gamma(3)) = \beta(3) = 3$. So, by Theorem 3.3, H is a subgroup. $|H| = 6$. The proof is valid for all $n \geq 3$. In the general case, $|H| = (n-2)!$. When S_n is replaced by A_n, $|H| = (n-2)!/2$.

45. $\langle(1234)\rangle$; $\{(1), (12), (34), (12)(34)\}$

46. α^k has k n/k-cycles.

47. This follows directly from Corollary 3 of Theorem 4.2.

48. Let $\alpha = (12)$ and $\beta = (13)$.

49. Let $\alpha = (123)$ and $\beta = (145)$.

50. $R_0 = (1)(2)(3)$; $R_{120} = (123)$; $R_{240} = (132)$; the reflections are (12), (13), (23).

51. Observe that (12) and (123) belong to S_n for all $n \geq 3$ and they do not commute. Observe that (123)(124) and (124)(123) belong to A_n for all $n \geq 4$ and they do not commute.

52. For any permutation α in S_7, α^2 is even, whereas (1234) is odd. For the second part note that because $\langle x^3\rangle$ is a subgroup of $\langle x\rangle$ and has order 4, we know 4 divides $|x|$. So we see from Example 5 that the possible values for $|x|$ are 4 and 12. If $|x| = 4$, then $(1234) = x^3 = x^{-1}$ and therefore $x = (4321)$. If $|x| = 12$, then $x = (4321)(567)$ or $x = (4321))(576)$ since no other choices yield $x^3 = (1234)$.

53. An even number of 2-cycles followed by an even number of 2-cycles gives an even number of 2-cycles in all. So the Finite Subgroup Test is verified.

54. Theorem 5.2 shows that disjoint cycles commute. For the other half, assume that $a = c$. Since $(ab) \neq (cd)$, we know $b \neq d$. If $(ab)(cd) = (cd)(ab)$, then $(ab)(cd) = (ab)(ad) = (adb)$ and $(cd)(ab) = (ad)(ab) = (abd)$. This means that a maps to both b and d, which are distinct. But permutations are 1-1 mappings.

55. Observe that $H = \{\beta \in S_n \mid \beta(\{1,2\}) = \{1,2\}$. So if $\alpha, \beta \in H$, then $(\alpha\beta)(\{1,2\}) = \alpha(\beta(\{1,2\}) = \alpha(\{1,2\}) = \{1,2\}$. So H is a subgroup. To find $|H|$, observe that for elements of H there are two choices for the image of 1, then no choice for the image of 2, and $(n-2)!$ choices for the remaining $n-2$ images. So, $|H| = 2(n-2)!$

56. Since $\alpha\beta = (1, 51, 52, \ldots, 100, 2, 3, \ldots, 49$, $|\alpha\beta| = 100$. So, $|\alpha^{-1}| = 100$, $|\alpha\beta| = 100$, and $|\alpha^{-1}\alpha\beta| = |\beta| = 3$.

57. R_0, R_{180}, H, V.

58. 216° rotation; reflection about the axis joining vertex 1 to the midpoint of the opposite side.

59. Labeling consecutive vertices of a regular 5-gon $1, 2, 3, 4, 5$, the even permutation $(14)(23)(5)$ is the reflection that fixes 5 and switches vertices 1 and 4 and 2 and 3. Multiplying the n rotations by this reflection yields all n reflections. There is no reflection in D_7 since their disjoint cycle form is a 1-cycle and three 2-cycles, which is an odd permutation.

60. Any element from A_n is expressible as a product of an even number of 2-cycles. For each pair of 2-cycles there are two cases. One is that they share an element in common $(ab)(ac)$, and the other is that they are disjoint $(ab)(cd)$. Now observe that $(ab)(ac) = (abc)$ and $(ab)(cd) = (cbd)(acb)$.

61. Since $(1234)^2$ is in B_n, it is non-empty. If $\alpha = \alpha_1\alpha_2\cdots\alpha_i$ and $\beta = \beta_1\beta_2\cdots\beta_j$ where i and j are even and all the α's and β's are 4-cycles, then $\alpha\beta = \alpha_1\alpha_2\cdots\alpha_i\beta_1\beta_2\cdots\beta_j$ is the product of $i + j$ 4-cycles and $i + j$ is even. So, by the finite subgroup test, B_n is a subgroup. To show that B_n is a subgroup of A_n, note that 4-cycles are odd permutations and the product of any two odd permutations is even. So, for the product of any even number of 4-cycles the product of the first two 4-cycles is even, then the product of the next two 4-cycles is even, and so on. This proves that B_n is a subgroup of A_n.

62. In Exercise 43 let G be A_5. Then stab(1) is the subgroup of A_5 consisting of the 24 even permutations of the set $\{2, 3, 4, 5\}$. Similarly, stab(2), stab(3), stab(4), stab(5) are subgroups of order 24.

63. The product of an element from $Z(A_4)$ of order 2 and an element of A_4 of order 3 would have order 6. But A_4 has no element of order 6.

64. $\alpha_1\alpha_2\alpha_3$ is odd and ϵ is even. The product of an odd number of 2-cycles cannot ϵ.

65. Cycle decomposition shows any nonidentity element of A_5 is a 5-cycle, a 3-cycle, or a product of a pair of disjoint 2-cycles. Then, observe there are $(5 \cdot 4 \cdot 3 \cdot 2 \cdot 1)/5 = 24$ group elements of the form $(abcde)$, $(5 \cdot 4 \cdot 3)/3 = 20$ group elements of the form (abc) and $(5 \cdot 4 \cdot 3 \cdot 2)/8 = 15$ group elements of the form $(ab)(cd)$. In this last case we must divide by 8 because there are 8 ways to write the same group element $(ab)(cd) = (ba)(cd) = (ab)(dc) = (ba)(dc) = (cd)(ab) = (cd)(ba) = (dc)(ab) = (dc)(ba)$.

66. One possibility for a cyclic subgroup is $\langle(1234)(5678)\rangle$. One possibility for a noncyclic subgroup is $\{(1), (12)(34), (56)(78), (12)(34)(56)(78)\}$.

67. If α has odd order n and α is an odd permutation, then $\epsilon = \alpha^n$ would be an odd permutation.

68. Using the notation in Table 5.1, α_2, α_3, and α_4 have order 2; $\alpha_5, \alpha_6, \ldots, \alpha_{12}$ have order 3. The orders of the elements divide the order of the group.

69. Since $|\alpha|$ is the least common multiple of the disjoint cycle lengths of α, let the distinct disjoint cycle lengths be α be $n_1, n_2, \ldots, n_k$. Because $n_1, n_2, \ldots, n_k$ are integers between 1 and n, each of them appears as a term in $n!$.

70. Suppose $\alpha \neq \varepsilon$ and $\alpha \in Z(S_n)$. Write α in disjoint cycle form $(a_1a_2\ldots)\ldots$, where $a_1 \neq a_2$. Let a_3 be different from a_1 and a_2 and let $\beta = (a_1)(a_2a_3)$. Then, $(\alpha\beta)(a_1) = a_2$, while $(\beta\alpha)(a_1) = a_3$.

71. That $a * \sigma(b) \neq b * \sigma(a)$ is done by examining all cases. To prove the general case, observe that $\sigma^i(a) * \sigma^{i+1}(b) \neq \sigma^i(b) * \sigma^{i+1}(a)$ can be written in the form

$\sigma^i(a) * \sigma(\sigma^i(b)) \neq \sigma^i(b) * \sigma(\sigma^i(a))$, which is the case already done. If a transposition were not detected, then
$\sigma(a_1)*\cdots*\sigma^i(a_i)*\sigma^{i+1}(a_{i+1})*\cdots*\sigma^n(a_n) = \sigma(a_1)*\cdots*\sigma^i(a_{i+1})*\sigma^{i+1}(a_i)*\cdots*\sigma^n(a_n)$,
which implies $\sigma^i(a_i) * \sigma^{i+1}(a_{i+1}) = \sigma^i(a_{i+1}) * \sigma^{i+1}(a_i)$.

72. 5

73. By Theorem 5.4 it is enough to prove that every 2-cycle can be expressed as a product of elements of the form $(1k)$. To this end observe that if $a \neq 1, b \neq 1$, then $(ab) = (1a)(1b)(1a)$.

74. By case-by-case, analysis H is a subgroup for $n = 1, 2, 3$, and 4. For $n \geq 5$, observe that (12)(34) and (12)(35) belong to H but their product does not.

75. ADVANCE WHEN READY

76. TAAKTPKSTOOPEDN

CHAPTER 6

Isomorphisms

1. Property 1: Because ϕ is 1-1 and onto, so is ϕ^{-1}. Moreover, $\phi^{-1}(\phi(a)\phi(b)) = \phi^{-1}(\phi(ab)) = ab = \phi^{-1}(\phi(a))\phi^{-1}(\phi(b))$.
Property 2: $\phi(e) = e$ is in $\phi(K)$, so $\phi(K)$ is not empty. Also, $\phi(a), \phi(b) \in \phi(K)$ implies that $\phi(a)\phi(b)^{-1} = \phi(a)\phi(b^{-1})\phi(ab^{-1} \in \phi(K)$.

2. An automorphism of a cyclic group must carry a generator to a generator. Thus $1 \to 1$ and $1 \to -1$ are the only two choices for the image of 1. So let $\alpha : n \to n$ and $\beta : n \to -n$. Then $\text{Aut}(Z) = \{\alpha, \beta\}$. The same is true for $\text{Aut}(Z_6)$.

3. ϕ is onto since any positive real number r is the image of $\sqrt{r}$. ϕ is one-to-one since $\sqrt{a} = \sqrt{b}$ implies that $a = b$. Finally, $\phi(xy) = \sqrt{xy} = \sqrt{x}\sqrt{y} = \phi(x)\phi(y)$.

4. $U(8)$ is not cyclic while $U(10)$ is. Define ϕ from $U(8)$ to $U(12)$ by $\phi(1) = 1$; $\phi(3) = 5$; $\phi(5) = 7$; $\phi(7) = 11$. To see that ϕ is operation preserving we observe that $\phi(1a) = \phi(a) = \phi(a) \cdot 1 = \phi(a)\phi(1)$ for all a; $\phi(3 \cdot 5) = \phi(7) = 11 = 5 \cdot 7 = \phi(3)\phi(5)$; $\phi(3 \cdot 7) = \phi(5) = 7 = 5 \cdot 11 = \phi(3)\phi(7)$; $\phi(5 \cdot 7) = \phi(3) = 5 = 7 \cdot 11 = \phi(5)\phi(7)$.

5. The mapping $\phi(x) = (3/2)x$ is an isomorphism from G onto H. Multiplication is not preserved. When $G = \langle m \rangle$ and $H = \langle n \rangle$ the mapping $\phi(x) = (n/m)x$ is an isomorphism from G onto H.

6. The identity is an isomorphism from G onto G. If β is an isomorphism from G onto H, then β^{-1} is an isomorphism from H onto G (see Theorem 6.3). If β is an isomorphism from G onto H, and α is an isomorphism from H onto K, then $\alpha\beta$ is an isomorphism from G onto K. That $\alpha\beta$ is one-to-one and onto is done in Theorem 0.8. If $a, b \in G$, then
$(\alpha\beta)(ab) = \alpha(\beta(ab)) = \alpha(\beta(a)\beta(b)) = \alpha(\beta(a))\alpha(\beta(b)) = (\alpha\beta)(a)(\alpha\beta)(b)$.

7. D_{12} has an element of order 12 and S_4 none; D_{12} has and element of order 6 and S_4 none; D_{12} has 2 elements of order 3 and S_4 has 8; D_{12} has 13 elements of order 2 and S_4 has 9.

8. Properties of real numbers assure that the mapping is one-to-one and onto **R** and that $\log_{10}(ab) = \log_{10}(a) + \log_{10}(b)$ is a property of logs.

9. Since $T_e(x) = ex = x$ for all x, T_e is the identity. For the second part, observe that $T_g \circ (T_g)^{-1} = T_e = T_{gg^{-1}} = T_g \circ T_{g^{-1}}$ and cancel.

10. $\phi(na) = n\phi(a)$.

11. $3\overline{a} - 2\overline{b}$.

12. Suppose α is an automorphism of G. Then, $\alpha(ab) = (ab)^{-1}$ and $\alpha(ab) = \alpha(a)\alpha(b) = a^{-1}b^{-1}$. So, $a^{-1}b^{-1} = (ab)^{-1} = b^{-1}a^{-1}$ for all a and b in G. Taking the inverse of both sides proves that G is Abelian.

    If G is Abelian, then for all a and b in G, we have $(ab)^{-1} = (ba)^{-1} = a^{-1}b^{-1}$. Thus, $\alpha(ab) = \alpha(a)\alpha(b)$.

    That α is one-to-one and onto follows directly from the definitions.

13. For any x in the group, we have
$(\phi_g\phi_h)(x) = \phi_g(\phi_h(x)) = \phi_g(hxh^{-1}) = ghxh^{-1}g^{-1} = (gh)x(gh)^{-1} = \phi_{gh}(x)$.

14. In $U(16)$, $|7| = |9| = 2$, so by Theorem 4.4, $U(16)$ is not cyclic.

15. ϕ_{R_0} and $\phi_{R_{90}}$ disagree on H; ϕ_{R_0} and ϕ_H disagree on R_{90}; ϕ_{R_0} and ϕ_D disagree on R_{90}; $\phi_{R_{90}}$ and ϕ_H disagree on R_{90}; $\phi_{R_{90}}$ and ϕ_D disagree on R_{90}; ϕ_H and ϕ_D disagree on D.

16. $\text{Aut}(Z_2) \approx \text{Aut}(Z_1) \approx Z_1$;
$\text{Aut}(Z_6) \approx \text{Aut}(Z_4) \approx \text{Aut}(Z_3) \approx U(6) \approx Z_2$;
$\text{Aut}(Z_{10}) \approx \text{Aut}(Z_5) \approx Z_4$ (see Example 4 and Theorem 6.4);
$\text{Aut}(Z_{12}) \approx \text{Aut}(Z_8)$ (see Exercise 4 and Theorem 6.4).

17. We must show that $\text{Aut}(G)$ has an identity, $\text{Aut}(G)$ is closed, the composition of automorphisms is associative, and the inverse of every element in $\text{Aut}(G)$ is in $\text{Aut}(G)$. Clearly, the identity function $\epsilon(x) = x$ is 1-1, onto, and operation preserving. For closure let $\alpha, \beta \in \text{Aut}(G)$. That $\alpha\beta$ is 1-1 and onto follows from Theorem 0.8. For $a, b \in G$, we have
$(\alpha\beta)(ab) = \alpha(\beta(ab)) = \alpha(\beta(a)\beta(b)) = (\alpha(\beta(a))(\alpha(\beta(b)) = (\alpha\beta)(a)(\alpha\beta)(b)$.
Associativity follows from properties of functions (see Theorem 0.8). Let $\alpha \in \text{Aut}(G)$. Theorem 0.8 shows that α^{-1} is 1-1 and onto. We must show that α^{-1} is operation preserving: $\alpha^{-1}(xy) = \alpha^{-1}(x)\alpha^{-1}(y)$ if and only if $\alpha(\alpha^{-1}(xy)) = \alpha(\alpha^{-1}(x)\alpha^{-1}(y))$. That is, if and only if $xy = \alpha(\alpha^{-1}(x))\alpha(\alpha^{-1}(y)) = xy$. So α^{-1} is operation preserving.

To prove that $\text{Inn}(G)$ is a group, we may use the subgroup test. Exercise 13 shows that $\text{Inn}(G)$ is closed. From $\phi_e = \phi_{gg^{-1}} = \phi_g\phi_{g^{-1}}$ we see that the inverse of ϕ_g is in $\text{Inn}(G)$. That $\text{Inn}(G)$ is a group follows from the equation $\phi_g\phi_h = \phi_{gh}$.

18. Let ϕ be an isomorphism from G to H. For any β in $\text{Aut}(G)$, define a mapping from $\text{Aut}(G)$ to $\text{Aut}(H)$ by $\Gamma(\beta) = \phi\beta\phi^{-1}$. Then, Γ is 1-1 and operation preserving. (See Theorem 0.8 and Exercise 6). To see that Γ is onto, observe that for any γ in $\text{Aut}(H)$, $\Gamma(\phi^{-1}\gamma\phi) = \gamma$.

19. Note that for $n > 1$, $(\phi_a)^n = (\phi_a)^{n-1}\phi_a$, so an induction argument gives $(\phi_a)^n = (\phi_a^{n-1})\phi_a = \phi_{a^{n-1}}\phi_a$. Thus $(\phi_{a^{n-1}}\phi_a)(x) = \phi_{a^{n-1}}(\phi_a(x)) = \phi_{a^{n-1}}(\phi_a)(x)) = \phi_{a^{n-1}}(axa^{-1}) = a^{n-1}(axa^{-1})(a^{n-1})^{-1} = a^{n-1}(axa^{-1})(a^{-n+1}) = a^nxa^{-n} = \phi_{a^n}(x)$.
To handle the case where n is negative, we note that
$\phi_e = \phi_{a^na^{-n}} = \phi_{a^n}\phi_{a^{-n}} = \phi_{a^n}(\phi_a)^{-n}$ (because $-n$ is positive). Solving for ϕ_{a^n} we obtain $\phi_{a^n}\phi_{a^{-n}} = \phi_{a^n} = (\phi_a)^n$.

20. This follows from Exercise 19.

21. Since $b = \phi(a) = a\phi(1)$ it follows that $\phi(1) = a^{-1}b$ and therefore $\phi(x) = a^{-1}bx$. (Here a^{-1} is the multiplicative inverse of a mod n, which exists because $a \in U(n)$.)

22. Note that ϕ must take $R_{360°/n}$ to an element of order n. Since the reflections have order 2, we know $\phi(R_{360°/n})$ is a rotation. Thus, $\phi(H) = \phi(\langle R_{360°/n}\rangle) \subseteq K$.

23. Note that both H and K are isomorphic to the group of all permutations of four symbols, which is isomorphic to S_4. The same is true when 5 is replaced by n since both H and K are isomorphic to S_{n-1}.

24. Observe that $\langle 2\rangle, \langle 3\rangle, \ldots$ are distinct and each is isomorphic to Z.

25. Recall when n is even, $Z(D_n) = \{R_0, R_{180}\}$. Since R_{180} and $\phi(R_{180})$ are not the identity and belong to $Z(D_n)$ they must be equal.

26. Note that $\phi(e) = e = e^4$ so H is not empty. Let a and b belong to H. Then, $\phi(ab^{-1}) = \phi(a)\phi(b^{-1}) = a^4(b^{-1})^4 = (ab^{-1})^4$. This proof works for any integer n. If $\phi(x) = x^{-1}$ and $a \in H$, then $\phi(a) = a^{-1} = a^4$, so $a^5 = e$. Thus $|a| = 1$ or 5.

27. Z_{60} contains cyclic subgroups of orders 12 and 20 and any cyclic group that has subgroups or orders 12 and 20 must be divisible by 12 and 20. So, 60 is the smallest order of any cyclic group that has subgroups isomorphic to Z_{12} and Z_{60}.

28. $\phi(5) = 5 \bmod 20$ is the same as $5\phi(1) = 5, 25, 45, 65, 85$ in Z. But we also need $|\phi(1)| = k = 20$. So, we need $\gcd(n, k) = 1$. This gives us $\phi(x) = x$; $\phi(x) = 9x$; $\phi(x) = 13x$; $\phi(x) = 17x$.

29. See Example 16 of Chapter 2.

30. Since 3 is an element of Z_{20} with order 4, the mapping $\phi(x) = 3x$ is in Aut(20) and applying it to 1 four times sends 1 to 1. So, for any k in Z_{20}, $\phi(k) = \phi(k \cdot 1) = k\phi(1) = k$.

31. That α is one-to-one follows from the fact that r^{-1} exists module n. Onto follows from Exercise 13 in Chapter 5. The operation preserving condition is Exercise 11 of Chapter 0.

32. The mapping $\begin{bmatrix} 1 & a \\ 0 & 1 \end{bmatrix} \to a$ is an isomorphism to Z when $a \in Z$ and to $\mathbf{R}$ when $a \in \mathbf{R}$.

33. By Part 1 of Theorem 6.2, we have $\phi(a^n) = \phi(a)^n = \gamma(a)^n = \gamma(a^n)$ thus ϕ and γ agree on all elements of $\langle a \rangle$.

34. Observe that $\phi(7) = 7\phi(1) = 13$ and since 7 is relatively prime to 50, 7^{-1} exists modulo 50. Thus, we have $\phi(1) = 7^{-1} \cdot 13 = 43 \cdot 13 = 9$ and $\phi(x) = \phi(x \cdot 1) = x\phi(1) = 9x$.

35. First observe that because $2^5 = 10 = -1$ we have $|2| = 10$. So, by parts 4 and 2 of Theorem 6.1, the mapping that takes $\phi(x) = 2^x$ is an isomorphism.

36. For all automorphisms ϕ of Q^* we know that $\phi(1) = 1$ and $\phi(-1) = -1$. For any rational $a/b = p_1^{m_1} p_2^{m_2} \cdots p_s^{m_s} / q_1^{n_1} q_2^{n_2} \cdots q_s^{n_t}$ we have $\phi(a/b) = \phi(p_1)^{m_1} \phi(p_2)^{m_2} \cdots \phi(p_s)^{m_s} \phi(q_1)^{-n_1} \phi(q_2)^{-n_2} \cdots \phi(q_s)^{-n_t}$.

37. $T_g(x) = T_g(y)$ if and only if $gx = gy$ or $x = y$. This shows that T_g is a one-to-one function. Let $y \in G$. Then $T_g(g^{-1}y) = y$, so that T_g is onto.

38. $U(20)$ has three elements of order 2, whereas $U(24)$ has seven.

39. To prove that ϕ is 1-1, observe that $\phi(a + bi) = \phi(c + di)$ implies that $a - bi = c - di$. From properties of complex numbers this gives that $a = c$ and $b = d$. Thus $a + bi = c + di$. To prove ϕ is onto, let $a + bi$ be any complex number. Then $\phi(a - bi) = a + bi$. To prove that ϕ preserves addition and multiplication, note that $\phi((a + bi) + (c + di)) = \phi((a + c) + (b + d)i) = (a + c) - (b + d)i = (a - bi) + (c - di) = \phi(a + bi) + \phi(c + di)$. Also, $\phi((a + bi)(c + di)) = \phi((ac - bd) + (ad + bc)i) = (ac - bd) - (ad + bc)i$ and $\phi(a + bi)\phi(c + di) = (a - bi)(c - di) = (ac - bd) - (ad + bc)i$.

40. Map $a + b\sqrt{2} \to \begin{bmatrix} a & 2b \\ b & a \end{bmatrix}$. This map preserves both addition and multiplication.

41. First observe that Z is a cyclic group generated by 1. By property 3 of Theorem 6.2, it suffices to show that Q is not cyclic under addition. By way of contradiction, suppose that $Q = \langle p/q \rangle$. But then $p/2q$ is a rational number that is not in $\langle p/q \rangle$.

42. S_8 contains $\langle (12345)(678) \rangle$ which has order 15. Since $|U(16)| = 8$, by Cayley's Theorem, S_8 contains a subgroup isomorphic to $U(16)$. The elements of D_8 can be represented as permutations of the 8 vertices of a regular 8-gon.

43. The notation itself suggests that

$$\phi(a+bi) = \begin{bmatrix} a & -b \\ b & a \end{bmatrix}$$

is the appropriate isomorphism. To verify this, note that

$$\phi((a+bi)+(c+di)) = \begin{bmatrix} a+c & -(b+d) \\ (b+d) & a+c \end{bmatrix} =$$

$$\begin{bmatrix} a & -b \\ b & a \end{bmatrix} + \begin{bmatrix} c & -d \\ d & c \end{bmatrix} = \phi(a+bi) + \phi(c+di).$$

Also, $\phi((a+bi)(c+di)) = \phi((ac-bd)+(ad+bc)i) =$

$$\begin{bmatrix} (ac-bd) & -(ad+bc)) \\ (ad+bc) & ac-bd) \end{bmatrix} = \begin{bmatrix} a & -b \\ b & a \end{bmatrix}\begin{bmatrix} c & -d \\ d & c \end{bmatrix} =$$

$$\phi(a+bi)\phi(c+di).$$

44. $\phi((a_1,\ldots,a_n)+(b_1,\ldots,b_n)) = (-a_1,\ldots,-a_n) = (-b_1,\ldots,-b_n)$ implies $(a_1,\ldots,a_n) = (b_1,\ldots,b_n))$ so that ϕ is 1-1. For any $(a_1,\ldots,a_n)$, we have $\phi(-a_1,\ldots,-a_n) = (a_1,\ldots,a_n)$ so ϕ is onto.
$\phi((a_1+b_1,\ldots,a_n+b_n)) = (-(a_1+b_1),\ldots,-(a_n+b_n)) =$
$(-a_1,\ldots,-a_n)(-b_1,\ldots,-b_n) = \phi((a_1,\ldots,a_n)) + \phi((b_1,\ldots,b_n))$.
ϕ reflects each point through the origin.

45. Yes, by Cayley's Theorem.

46. $(ab)^2 = a^2b^2$ shows that the mapping is O.P. To show it is 1-1, note that $a^2 = b^2$, implies $e = a^2b^{-2} = (ab^{-1})^2$ so that $|ab^{-1}| = 1$ or 2. Thus, $a = b$. Since G is finite, 1-1 implies onto. For Z under addition, $g \to 2g$ is not onto but is $1-1$ and operation preserving.

47. Observe that $\phi_g(y) = gyg^{-1}$ and $\phi_{zg}(y) = zgy(zg)^{-1} = zgyg^{-1}z^{-1} = gyg^{-1}$, since $z \in Z(G)$. So, $\phi_g = \phi_{zg}$.

48. In **R** under addition every nonzero element has infinite order. In $\mathbf{R}^*$ under multiplication -1 has order 2.

49. $\phi_g = \phi_h$ implies $gxg^{-1} = hxh^{-1}$ for all x. This implies $h^{-1}gx(h^{-1}g)^{-1}) = x$, and therefore $h^{-1}g \in Z(G)$. $\phi_g = \phi_h$ if and only if $h^{-1}g \in Z(G)$.

50. $\alpha(x) = (12)x(12)$ and $\beta(x) = (123)x(123)^{-1}$.

51. By Exercise 49, $\phi_\alpha = \phi_\beta$ implies $\beta^{-1}\alpha$ is in $Z(S_n)$ and by Exercise 70 in Chapter 5, $Z(S_n) = \{\epsilon\}$, which implies that $\alpha = \beta$.

52. Since 2 is not in the image, the mapping is not onto.

53. Since both ϕ and γ take e to itself, H is not empty. Assume a and b belong to H. Then $\phi(ab^{-1}) = \phi(a)\phi(b^{-1}) = \phi(a)\phi(b)^{-1} = \gamma(a)\gamma(b)^{-1} = \gamma(a)\gamma(b^{-1}) = \gamma(ab^{-1})$. Thus ab^{-1} is in H.

54. G is Abelian.

55. $(12)H(12)$ and $(123)H(123)^{-1}$.

56. Since $|R_{45}| = 8$, it must map to elements of order 8. Since the integers between 1 and 8 relatively prime to 8 are 1, 3, 5, 7, the elements of order 8 are $R_{45}, {R_{45}}^3, {R_{45}}^5, {R_{45}}^7$.

57. Since -1 is the unique element of $\mathbf{C}^*$ of order 2, $\phi(-1) = -1$. Since i and $-i$ are the only elements of $\mathbf{C}^*$ of order 4, $\phi(i) = i$ or $-i$.

58. If a is an element not in $Z(G)$, then ϕ_a is not the identity.

59. Z_{120}, D_{60}, S_5. Z_{120} is Abelian, the other two are not. D_{60} has an element of order 60 and S_5 does not.

60. Using Exercise 25 we have $\phi(V) = \phi(R_{180}H) = \phi(R_{180})\phi(H) = R_{180}D = D'$.

61. For the first part, observe that $\phi(D) = \phi(R_{90}V) = \phi(R_{90})\phi(V) = R_{270}V = D'$ and $\phi(H) = \phi(R_{90}D) = \phi(R_{90})\phi(D) = R_{270}D' = H$. For the second part, we have that $\phi(D) = \phi(R_{90}V) = \phi(R_{90})\phi(V) = R_{90V} = D$ and $\phi(H) = \phi(R_{180}V) = (\phi(R_{90}))^2\phi(V) = {R_{90}}^2V = R_{180}V = H$.

62. $\alpha_5 = (0)(157842)(36); \alpha_8 = (0)(18)(27)(36)(45)$.

63. $(R_0R_{90}R_{180}R_{270})(HD'VD)$.

64. In the general case, because ϕ is 1-1, G has no element whose order divides order n.

65. The first statement follows from the fact that every element of D_n has the form $R^i_{360/n}$ or $R^i_{360/n}F$. Because α must map an element of order n to an element of order n, $R_{360/n}$ must map to $R^i_{360/n}$ where $i \in U(n)$. Moreover, F must map to a reflection (see Exercise 20). Thus we have at most $n|U(n)|$ choices.

66. By the previous exercise, the mapping from D_n to $\text{Inn}(D_n)$ that takes a to the inner automorphism induced by $\phi_a(x) = axa^{-1}$ is 1-1 when n is odd, and 2-1 when n is even.

67. In both cases H is isomorphic to the set of all even permutations of the set of four integers, so it is isomorphic to A_4.

68. By part 2 of Theorem 6.1, $n\phi(1/n) = \phi(1)$ so that $\phi(1/n) = (1/n)\phi(1)$. Also, by Part 2 of Theorem 6.2, $\phi(m/n) = m\phi(1/n) = m(1/n)\phi(1) = (m/n)\phi(1)$.

69. The mapping $\phi(x) = x^2$ is one-to-one from Q^+ to Q^+ since $a^2 = b^2$ implies $a = b$ when both a and b are positive. Moreover, $\phi(ab) = \phi(a)\phi(b)$ for all a and b. However, ϕ is not onto since there is no rational whose square is 2. So, the image of ϕ is a proper subgroup of Q^+.

70. The argument given in Exercise 68 shows that an isomorphic image of Q has the form aQ where $a = \phi(1)$. But $aQ = Q$.

71. Suppose that ϕ is an automorphism of R^* and a is positive. Then $\phi(a) = \phi(\sqrt{a}\sqrt{a}) = \phi(\sqrt{a})\phi(\sqrt{a}) = \phi(\sqrt{a})^2 > 0$. Now suppose that a is negative but $\phi(a) = b$ is positive. Then, by the case we just did, $a = \phi^{-1}(\phi(a)) = \phi^{-1}(b)$ is positive. This is a contradiction.
Here is an alternate argument for the case that a is negative and $\phi(a)$ is positive. Because -1 is the only real number of order 2 and the first case, we know that $0 < \phi(-a) = \phi(-1)\phi(a) = -\phi(a)$, which is a contradiction.

72. If ϕ were an isomorphism, then $0 = \phi(1) = \phi(-1 \cdot -1) = \phi(-1) + \phi(-1) = 2\phi(-1)$ implies that $\phi(-1) = 0$. But then, ϕ is not $1-1$.

73. Say ϕ is an isomorphism from Q to $\mathbf{R}^+$ and ϕ takes 1 to a. It follows that the integer r maps to a^r. Then $a = \phi(1) = \phi(s\frac{1}{s}) = \phi(\frac{1}{s} + \cdots + \frac{1}{s}) = \phi(\frac{1}{s})^s$ and therefore $a^{\frac{1}{s}} = \phi(\frac{1}{s})$. Thus, the rational r/s maps to $a^{r/s}$. But $a^{r/s} \neq a^{\pi}$ for any rational number r/s.

74. In Q^* $|-1| = 2$, whereas the only element of finite order in Q^+ is 1.

75. Send each even permutation in S_n to itself. Send each odd permutation α in S_n to $\alpha(n+1, n+1)$. This does not contradict Theorem 5.5, because the subgroup is merely isomorphic to A_{n+2}, not the same as A_{n+2}. In particular, this example shows that an isomorphism from one permutation group to another permutation group need not preserve oddness.

CHAPTER 7

Cosets and Lagrange's Theorem

1. $H, 1+H, 2+H$. To see that there are no others, notice that for any integer n we can write $n = 3q + r$ where $0 \leq r < 3$. So, $n + H = r + 3q + H = r + H$, where $r = 0, 1$ or 2. For the second part there are n left cosets: $0 + \langle n \rangle, 1 + \langle n \rangle, \ldots, n - 1 + \langle n \rangle$.

2. $b - a \in H$.

3.

a. $11 + H = 17 + H$ because $17 - 11 = 6$ is in H;

b. $-1 + H = 5 + H$ because $5 - (-1) = 6$ is in H;

c. $7 + H \neq 23 + H$ because $23 - 7 = 16$ is not in H.

4. Since $8/2 = 4$, there are four cosets. Let $H = \{1, 11\}$. The cosets are $H, 7H, 13H, 19H$.

5. Five: $\langle a^5 \rangle, a\langle a^5 \rangle, a^2\langle a^5 \rangle, a^3\langle a^5 \rangle, a^4\langle a^5 \rangle$. Since $\langle a^4 \rangle = \langle a^2 \rangle$ there are two cosets: $\langle a^4 \rangle, a\langle a^4 \rangle$.

6. Let F and F' be distinct reflections in D_3. Then, take $H = \{R_0, F\}$ and $K = \{R_0, F'\}$.

7. Suppose that $H \neq \langle 3 \rangle$. Let $a \in H$ but a not in $\langle 3 \rangle$. Then $a + \langle 3 \rangle = 1 + \langle 3 \rangle$ or $a + \langle 3 \rangle = 2 + \langle 3 \rangle$.

8. Observe that $aH = bK$ implies that $b^{-1}aH = K$. Thus, $b^{-1}aH$ is a subgroup of G. From part 9 of the lemma in Chapter 7, we have that $b^{-1}a$ is in H and therefore $b^{-1}aH = H$.

9. Let ga belong to $g(H \cap K)$ where a is in $H \cap K$. Then by definition, ga is in $gH \cap gK$. Now let $x \in gH \cap gK$. Then $x = gh$ for some $h \in H$ and $x = gk$ for some $k \in K$. Cancellation then gives $h = k$. Thus $x \in g(H \cap K)$.

10. By Lagrange's Theorem, $|H| = 5, 31$ or 155. But $|H| = 5$ implies that all non-identity elements in H have order 5 and $|H| = 31$ implies that all non-identity elements in H have order 31.

11. Suppose that $h \in H$ and $h < 0$. Then $h\mathbf{R}^+ \subseteq hH = H$. But $h\mathbf{R}^+$ is the set of all negative real numbers. Thus $H = \mathbf{R}^*$.

12. The coset containing $c + di$ is the circle with center at the origin and radius $\sqrt{c^2 + d^2}$.

13. By Lagrange's Theorem the possible orders are 1, 2, 3, 4, 5, 6, 10, 12, 15, 20, 30, 60.

14. 84 or 210.

15. By Lagrange's Theorem, the only possible orders for the subgroups are 1, p and q. By Corollary 3 of Lagrange's Theorem, groups of prime order are cyclic. The subgroup of order 1 is $\langle e \rangle$.

16. Note that $\phi(n) = |U(n)|$, then use Corollary 4 of Lagrange's Theorem and mimic the proof of Corollary 5 of Lagrange's Theorem.

17. By Exercise 16 we have $5^6 \bmod 7 = 1$. So, using mod 7 we have $5^{15} = 5^6 \cdot 5^6 \cdot 5^2 \cdot 5 = 1 \cdot 1 \cdot 4 \cdot 5 = 6$; $7^{13} \bmod 11 = 2$.

18. Note that $n - 1 \in U(n)$ and has order 2.

19. By Corollary 4 of Theorem 7.1, $g^n = e$. Then since $g^m = e$ and $g^n = e$ we know by Corollary 2 of Theorem 4.1 that $|g|$ is a common divisor of both m and n. So, $|g| = 1$.

20. Since $|H \cap K|$ must divide 12 and 35, $|H \cap K| = 1$. If H and K are relatively prime, $|H \cap K| = 1$.

21. First observe that for all $n \geq 3$ the subgroup of rotations of D_n is isomorphic to Z_n. If n is even let F be any reflection in D_n. Then the set $\{R_0, R_{180}, F, FR_{180}\}$ is closed and therefore a subgroup of order 4. Now suppose that D_n has a subgroup K of order 4. By Lagrange, $|D_n| = 2n = 4k$ and therefore $n = 2k$.

22. Since $|H \cap K|$ must divide both 12 and 18, it must divide 6. The case where $|H \cap K| = 1$ is trivial. The case where $|H \cap K| = p$ follows from Corollary 3 of Theorem 7.1. The case where $|H \cap K| = 2p$ follows from Theorem 7.3.

23. Since G has odd order, no element can have order 2. Thus, for each $x \neq e$, we know that $x \neq x^{-1}$. So, because G is Abelian, we can write the product of all the elements in the form $ea_1a_1^{-1}a_2a_2^{-1} \cdots a_na_n^{-1} = e$.

24. Let G be a group of order 4. If G has an element of order 4, then G is cyclic. So, every element in G has order 1 or 2. Then, for all $a, b \in G$, we have $ab = (ab)^{-1} = b^{-1}a^{-1} = ba$.

25. For $|G| = p^n$, the group is cyclic or $a^{p^{n-1}} = e$ for all a in G.

26. Since a is not in $Z(G)$ we know $C(a)$ is a proper subgroup of G. Thus, $|C(a)| \leq |G|/2$. But then the numbers of elements not in $C(a)$ is at least $|G|/2$. If G has odd order, then at least 2/3 of the elements in G do not commute with a.

27. Let G be a group of order 77. If G is cyclic, then by Theorem 4.3, G has exactly one subgroup of order 7 and exactly one of order 11. If G is not cyclic, by Lagrange, its 76 non-identity elements have orders 7 or 11. By Corollary of Theorem 4.4, G has at most 72 elements of order 6 and therefore must have elements of order 11, which come 10 at a time. Theorem 7.2 shows that G cannot have more that one subgroup of order 11, so there are exactly 66 elements of order 7.

28. Since G is non-Abelian, it has no element of order 8. By Lagrange, any nonidentity element must have order 2 or 4. Finally, by Exercise 45 of Chapter 2, not all nonidentity elements have order 2.

29. If the group is cyclic, Theorem 4.3 says that it has exactly one subgroup of order 5. So, assume the group is not cyclic. Not all of the 54 nonidentity elements can have order 5 because the number of elements of order 5 is a multiple of $\phi(5) = 4$. So the group has an element of order 11. Also, since $\phi(11) = 10$, the number of elements of order 11 is a multiple of 10. If there were more than 10, the group would have distinct subgroups H and K of order 11. But then $|HK| = |H||K|/|H \cap K| = 121$. So, excluding the subgroup of order 11, there are 44 elements remaining and each has order 5. That gives us exactly 11 subgroups of order 5.

30. Suppose that G has no element of order 105. Then, the possible orders of the elements are 1, 3, 5, 7, 15, 21 and 35. Since there is a unique subgroup of each of these orders, all the elements of order 3 occur in the unique subgroup of order 3, and likewise for the other orders. Now note that for any order k in this problem the number of elements of order k is $\phi(k)$. So, there is 1 element of order 1, 2 of order 3, 4 of order 5, 6 of order 7, 8 of order 15, 12 of order 21, and 24 of order 35. But this totals to fewer that 105, which is a contradiction.

    Alternate proof: Again, assume that there is no element of order 105. Let H be the unique subgroup of order 35 and K the unique subgroup of order 15. Observe that H must have an element of order 35, for otherwise all 34 non-identity elements of H would have order 5 or 7. But G has only 4 elements of order 5 and 6 of order 7. So, H contains all the elements of G of orders 1, 5, 7 and 35. Likewise, K contains all the elements of G of orders 1, 3, 5 and 15. Thus, $H \cup K$ contains every element of G. But $H \cup K$ has less than 51 elements (the exact number is 45, since the elements of orders 1 and 5 appear in both H and K).

31. By Lagrange's theorem every element in G has an order that is a divisor of n. So, we can partition the n elements of G according to their orders. For each divisor d of n, let m_d be the number of elements in G of order d. By our assumption, m_d is $\phi(d)$ where ϕ is the Euler phi function. (If there were more than $\phi(d)$ elements of order d in G, then G would have at least 2 subgroups of order d.) So $n = \Sigma m_d$ where d ranges over all divisors d of n. We also have from Exercise 87 of Chapter 4 that $n = \Sigma\phi(d)$ where d ranges over all divisors d of n. This proves that each $n_d = \phi(d)$. In particular, $m_n \neq 0$.

32. By Theorem 7.3, G is isomorphic to Z_{2p} or D_p. So the number of elements of order 2 is 1 or $p+1$.

33. Suppose that H and K are distinct subgroups of order m. Then $|HK| = |H||K| = \frac{m \cdot m}{|H \cap K|} \leq 2m$ and therefore $\frac{m}{2} \leq |H \cap K|$. Since m is odd and H and K are distinct, we know that $\frac{m}{2} < |H \cap K| < m$ and that $|H \cap K|$ divides m. This is impossible.

34. For any positive integer n let $\omega_n = \cos\left(\frac{2\pi}{n}\right) + i \sin\left(\frac{2\pi}{n}\right)$. The finite subgroups of C^* are those of the form $\langle \omega_n \rangle$. To verify this, let H denote any finite subgroup of C^* of order n. Then every element of H is a solution to $x^n = 1$. But the solution set of $x^n = 1$ in C^* is $\langle \omega_n \rangle$.

35. Observe that $|G : H| = |G|/|H|$, $|G : K| = |G|/|K|$, $|K : H| = |K|/|H|$. So, $|G : K||K : H| = |G|/|H| = |G : H|$.

36. By Exercise 35, $p = [G : H] = [G : K][K : H]$ so $[G : H] = 1$ or $[K : H] = 1$.

37. Cyclic subgroups of order 12 and 20 in D_n must be in the subgroup of rotations. So, n must be the smallest positive integer divisible by 12 and 20, which is 60.

38. Since $|g|$ must divide both 14 and 21, $|g| = 1$ or 7.

39. Let a have order 3 and b be an element of order 3 not in $\langle a \rangle$. Then $\langle a \rangle\langle b \rangle = \{a^i b^j \mid i = 0, 1, 2;\ j = 0, 1, 2\}$ is a subgroup of G of order 9. Now use Lagrange's Theorem.

40. Suppose that $|G|/|Z(G| = p$ where p is prime. Let $a \in G$, but $a \notin Z(G)$. Since the subgroup $C(a)$ contains both a and $Z(G)$ we know that $Z(G)$ is proper subgroup of

$C(a)$. Since $p = (|G|/|C(a)|)(|C(a)|/|Z(G)|)$ we have $|G|/|C(a)| = 1$ and therefore $C(a) = G$ and $a \in Z(G)$.

41. By Corollary 5 of Theorem 7.1, the statement is true for $n = 1$. For the sake of induction assume that $a^{p^k} = a$. Then $a^{p^{k+1}} = a^{p^k} a^p = a^p = a$.

42. Suppose G is a group of order 63. Let a be any non-identity element in G. By Lagrange's Theorem, $|a| = 63, 21, 9, 7$, or 3. If $|a| = 3$, we are done. If $|a| = 63$, then $|a^{21}| = 3$; if $|a| = 21$, then $|a^7| = 3$; and if $|a| = 9$, then $|a^3| = 3$. So, if any of these cases occur we are done. Thus, we may assume that all 62 non-identity elements in G have order 7. But, by the Corollary to Theorem 4.4, the number of elements of order 7 must be a multiple of 6.

43. Let $a \in G$ and $|a| = 5$. Then by Theorem 7.2 we know that the set $\langle a \rangle H$ has exactly $5 \cdot |H|/|\langle a \rangle \cap H|$ elements and $|\langle a \rangle \cap H|$ divides $|\langle a \rangle| = 5$. It follows that $|\langle a \rangle \cap H| = 5$ and therefore $\langle a \rangle \cap H = \langle a \rangle$.

44. Suppose the $a^k = b^k$. By the Corollary to Theorem 0.2, there are integers s and t such that $1 = ns + kt$. Then, by Corollary 4 of Lagrange's Theorem ,we have $a = a^{ns+kt} = (a^n)^s(a^k)^t = (a^k)^t = (b^n)^s(b^k)^t = b^{ns+kt} = b$. To prove that the mapping is an automorphism when the group is also Abelian, note that by Exercise 10 of Chapter 5 a 1-1 mapping from a finite set to itself is onto. Lastly, observe that $(ab)^k = a^k b^k$.

45. First observe that by Corollary 2 of Lagrange's Theorem every positive integer k with the property that $x^k = e$ for all x in G is a common multiple of orders of all the elements in G. So, d is the least common multiple of the orders of the elements of G. Since $|G|$ is a common multiple of the orders of all the elements of G it follows directly from the division algorithm (Theorem 0.1) that $|G|$ is divisible by d.

46. Since reflections have order 2, the subgroup must consist entirely of rotations and the subgroup of all rotations is cyclic.

47. Let G be a finite Abelian group. The case when G has 0 or 1 element of order 2 corresponds to the cases $n = 0$ and $n = 1$. Let $H = \{x \in G \mid x^2 = e\}$. Then H is a subgroup of G that consists of the identity and all elements of order 2. It suffices to prove $|H| = 2^n$. If G has at least two elements of order 2, say a_1 and b, then $H_1 = \{e, a_1, b, a_1 b\}$ is a subgroup of order 2^2. If $H_1 = H$, we are done. If not, let $a_2 \in H$ but not in H_1. Then $H_2 = H_1 \cup \langle a_2 \rangle H_1$ is a subgroup of H of order $2|H_1| = 2^3$. If $H_2 = H$ we are done. If not, let $a_3 \in H$ but not in H_2 and let $H_3 = H_2 \cup \langle a_3 \rangle H_2$. We can continue this argument until we reach H.

48. Suppose that an Abelian group G has order 15 and is not cyclic. Then, G has 14 elements of order 3 or 5. Since 14 is not a multiple $\phi(5) = 4$, G must have an element a of order 3. It follows from Example 7 of Chapter 3 and Theorem 7.2 that G cannot have more than one subgroup of order 3 or 5. Since the union of the unique subgroups of order 3 and 5 has only 7 elements, the remaining 8 must have order 15.

49. Let $|G| = 2k + 1$. Observe $a = ae = aa^{2k+1} = a^{2k+2} = (a^{k+1})^2$. To prove uniqueness suppose the $x^2 = a = y^2$. Then $x(x^2)^k = x^{2k+1} = e = y^{2k+1} = y(y^2)^k$. So, we can cancel the terms $(x^2)^k$ and $(y^2)^k$ to obtain $x = y$.

50. Let G be Abelian and $|G| = 4p^n$, where p is a prime. If G has an element a with $|a|$ of the form pm, then $|a^m| = p$. So, we may assume that all nonidentity elements of G have order 2 or 4. If G has an element a of order 4, then for any element b not in

$\langle a \rangle$, we have by Theorem 7.2 that $\langle a \rangle\langle b \rangle$ is a subgroup of G of order 8 or 16, which contradicts Lagrange's Theorem. Lastly, we may assume that every non-identity element of G has order 2. In this case, let a and b be distinct elements of G of order 2. Then $H = \{e, a, b, ab\}$ is a subgroup of G of order 4 and for any element c in G that is not in H we have $H\langle c \rangle$ is a subgroup of order 8.

51. If H and K are distinct subgroups of order p^m, then $np^m = |G| \geq |HK| = |H||K|/|H \cap K| \geq p^m p^m / p^{m-1} = pp^m$, which is obviously false.

52. If G has an element of order q^k or pq^k, then by the Fundamental Theorem of Cyclic Groups, it has an element of order q. So we may assume that every nonidentity element of G has order p. Let $a \neq e$ belong to G and b belong to G with b not in $\langle a \rangle$. Then, since $\langle a \rangle \cap \langle b \rangle = \{e\}$ we have by Theorem 7.2, that $pq^n = |G| \geq |\langle a \rangle\langle b \rangle|/|\langle a \rangle \cap \langle b \rangle| = p^2$. But then $q^n \geq p$.

53. Let G be the group and H the unique subgroup of order q. We must show that G has an element of order pq. Let a belong to G but not in H. By Lagrange, $|a| = p$ or pq. If $|a| = pq$ we are done. So, we may assume that a has order p and we let $K = \langle a \rangle$. Then $H \cup K$ accounts for $q + p - 1$ elements (the identity appears twice). Pick $b \in G$ but b not one of the elements in $H \cup K$. Then $L = \langle b \rangle$ is a subgroup of G of order p different than K. Then $K \cap L = \{e\}$ because $|K \cap L|$ must divide p and is not p. By Theorem 7.2 $|KL| = |K||L|/|K \cap L| = (p \cdot p)/1 = p^2$. But a group of order pq with $q < p$ cannot have p^2 elements. This shows that b cannot have order p. So $|b| = pq$.

54. $|H| = 1$ or p where p is a prime. To see this, suppose that $|H| > 1$ and let $a \in H$ and $a \neq e$. Let $|a| = pm$ where p is a prime. Then $|\langle a^m \rangle| = p$ and $H \subseteq \langle a^m \rangle$, so $|H| = p$. An example, where $|H| = p$, is $G = Z_{p^k}$ where p is prime and $k \geq 1$.

55. Let H and K be distinct subgroups of order 5. Then by Theorem 7.2 the subset HK has order 25. In the statement of the exercise, replace 5 with any prime p and 25 by p^2.

56. Say $|a| = n$. Then ${\phi_a}^n(x) = a^n x a^{-n} = x$, so that ${\phi_a}^n$ is the identity. For example, take $a = R_{90}$ in D_4. Then $|\phi_a| = 2$.

57. Since the order of G is divisible by both 10 and 25, it must be divisible by 50. But the only number less than 100 that is divisible by 50 is 50.

58. If $|Z(A_4)| > 1$, then A_4 would have an element of order 2 or order 3 that commutes with every element. But any subgroup generated by an element of order 2 and an element of order 3 that commute has order 6. This contradicts the fact shown in Example 5 that A_4 has no subgroup of order 6.

59. Let K be the set of all even permutations in H. Since K is closed, it is a subgroup of H. If $K = H$, we are done. If not, let α be an element in H that is odd. Then αK must be the set of all odd permutations in H, for if β is any odd permutation in H, we have $\alpha^{-1}\beta \in H$, which means $\beta \in H$. Thus $|H| = |K \cup \alpha K| = 2|K|$.

60. Observe that $|\text{orb}_{A_5}(5)| = 5$. Now use the Orbit-Stabilizer Theorem to show that $|\text{stab}_{A_5}(5)| = 12$. Note that the same argument applies to $\text{stab}_{A_5}(i)$ for $i = 1, 2, 3,$ and 4.

61. Suppose that H is a subgroup of A_5 of order 30. We claim that H contains all 20 elements of A_5 that have order 3. To verify this, assume that there is some α in A_5 of order 3 that is not in H. Then $A_5 = H \cup \alpha H$. It follows that $\alpha^2 H = H$ or

$\alpha^2 = \alpha H$. Since the latter implies that $\alpha \in H$, we have that $\alpha^2 H = H$, which implies that $\alpha^2 \in H$. But then $\langle\alpha\rangle = \langle\alpha^2\rangle \subseteq H$, which is a contradiction of our assumption that α is not in H. The same argument shows that H must contain all 24 elements of order 5. Since $|H| = 30$ we have a contradiction.

62. Suppose that H is a subgroup of A_5 of order 20. We claim that H contains all 24 elements of A_5 that have order 5. To verify this, assume that there is some α in A_5 of order 5 that is not in H. Then $A_5 = H \cup \alpha H \cup \alpha^2 H$. To see that the coset $\alpha^2 H$ is not the same as H, note that $\alpha^2 H = H$ implies that $\langle\alpha^2\rangle \subset H$ and $\langle\alpha\rangle = \langle\alpha^2\rangle$. Moreover, $\alpha^2 H$ is not the same as αH, for then $\alpha \in H$. It follows that $\alpha^3 H$ is equal to one of the cosets H, αH or $\alpha^2 H$. If $\alpha^3 H = H$, then $\alpha^3 \in H$ and therefore $\langle\alpha\rangle = \langle\alpha^3\rangle \subseteq H$, which contradicts the assumption that α is not in H. If $\alpha^3 H = \alpha H$, then $\alpha^2 \in H$ and therefore $\langle\alpha\rangle = \langle\alpha^2\rangle \subseteq H$, which contradicts the assumption that α is not in H. If $\alpha^3 H = \alpha^2 H$, then $\alpha \in H$, which contradicts the assumption that α is not in H. The same argument shows that H must contain all 24 elements of order 5. Since $|H| = 20$, we have a contradiction. An analogous argument shows that A_5 has no subgroup of order 15.

63. Say H is a subgroup of order 30. By Exercise 61, H is not a subgroup of A_5 and by Exercise 27 of Chapter 6, $H \cap A_5$ is a subgroup of A_5 of order 15. But this contradicts Exercise 52.

64. For any $\beta \in H$ we have $(\alpha\beta)(1)\alpha(\beta(1)) = \alpha(1) = 3$. So, αH is a subset of the set all elements of S_n that send 1 to 3. Let $\gamma(1) = 3$. Then $\alpha^{-1}\gamma(1) = \alpha^{-1}(3) = 1$. So, $\alpha^{-1}\gamma \in H$ and therefore $\gamma \in \alpha H$. In the general case, for $\alpha \in S_n$ $(n \geq 3)$ and i and j are integers between 1 and n with $\alpha(i) = j$ and $H = \text{stab}(i)$, then αH is a subset of the set all elements of S_n that send i to j.

65. If H is a subgroup of S_5 of order 60 other than A_5, then it follows from Theorem 7.2 that $|A_5 \cap H| = 30$, which contradicts Exercise 61.

66. $n = 1, 2, 3$. To see that there are no others, note that $|S_4| = 24$ does not divide 120, S_5 does not have an element of order 60 and D_{60} does, and for $n > 5$, $|S_n| > 120$.

67. Certainly, $a \in \text{orb}_G(a)$. Now suppose $c \in \text{orb}_G(a) \cap \text{orb}_G(b)$. Then $c = \alpha(a)$ and $c = \beta(b)$ for some α and β, and therefore $(\beta^{-1}\alpha)(a) = \beta^{-1}(\alpha(a)) = \beta^{-1}(c) = b$. So, if $x \in \text{orb}_G(b)$, then $x = \gamma(b) = \gamma(\beta^{-1}\alpha)(a)) = (\gamma\beta^{-1}\alpha)(a)$. This proves $\text{orb}_G(b) \subseteq \text{orb}_G(a)$. By symmetry, $\text{orb}_G(a) \subseteq \text{orb}_G(b)$.

68.

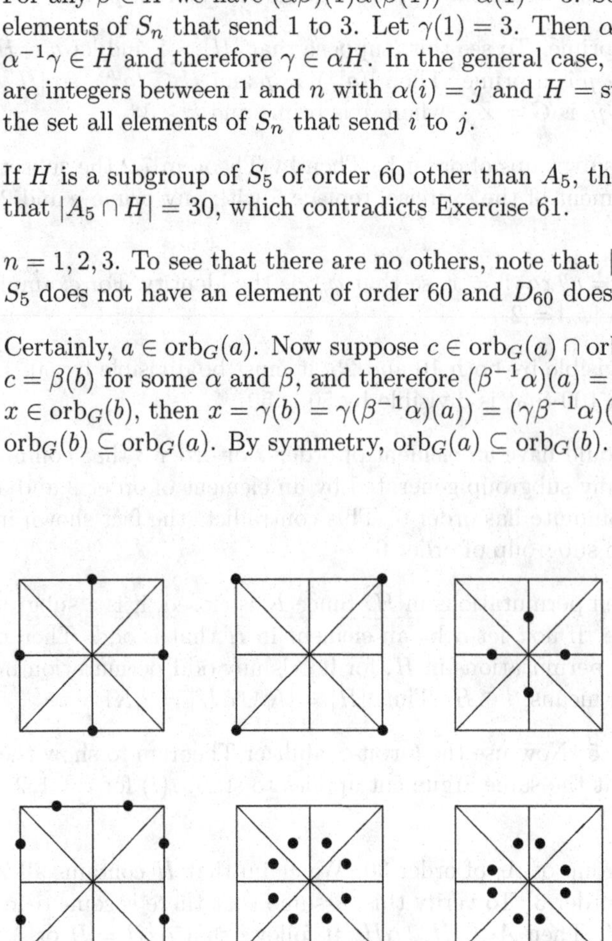

$\{R_0, H\}$; $\{R_0, D'\}$; $\{R_0, H\}$
$\{R_0\}$; $\{R_0\}$; $\{R_0\}$.

69. **a.** $\text{stab}_G(1) = \{(1), (24)(56)\}$; $\text{orb}_G(1) = \{1, 2, 3, 4\}$

b. $\text{stab}_G(3) = \{(1), (24)(56)\}$; $\text{orb}_G(3) = \{3, 4, 1, 2\}$

c. $\text{stab}_G(5) = \{(1), (12)(34), (13)(24), (14)(23)\}$; $\text{orb}_G(5) = \{5, 6\}$

70. Think of a cube as sitting on a table top with one face perpendicular to your line of sight. The four lines that join the upper corner of a cube to the midpoint of diametrically opposite vertical edge are axes of rotational symmetry of 120 degrees. The four lines that join the upper corner of a cube to the midpoint of lower horizontal edge at the maximum distance from the starting corner are also axes of rotational symmetry of 120 degrees.

71. Suppose that $B \in G$ and $\det B = 2$. Then $\det (A^{-1}B) = 1$, so that $A^{-1}B \in H$ and therefore $B \in AH$. Conversely, for any $Ah \in AH$ we have $\det Ah = (\det (A))(\det (h)) = 2 \cdot 1 = 2$.

72. The circle passing through Q, with center at P.

73. It is the set of all permutations that carry face 2 to face 1.

74. The order of the symmetry group would have to be $6 \cdot 20 = 120$.

75. If $aH = bH$, then $b^{-1}a \in H$. So $\det (b^{-1}a) = (\det b^{-1})(\det a) = (\det b^{-1})(\det a) = (\det b)^{-1}(\det a) = 1$. Thus $\det a = \det b$. Conversely, we can read this argument backwards to get that $\det a = \det b$ implies $aH = bH$.

76. **a.** 12 **b.** 24 **c.** 60 **d.** 60

77. To prove that the set is closed, note that $\alpha\beta^2 = (13) = \beta^2\alpha^3$, $\alpha^2\beta^2 = (14)(23) = \beta^2\alpha^2$, and $\alpha^3\beta^2 = (24) = \beta^2\alpha$.

CHAPTER 8

External Direct Products

1. Closure and associativity in the product follows from the closure and associativity in each component. The identity in the product is the n-tuple with the identity in each component. The inverse of $(g_1, g_2, \ldots, g_n)$ is $(g_1^{-1}, g_2^{-1}, \ldots, g_n^{-1})$.

2. In general, $(1, 1, \ldots, 1)$ is an element of largest order in $Z_{n_1} \oplus Z_{n_2} \oplus \cdots \oplus Z_{n_t}$. To see this, note that because the order of the 1 in each component is the order of the group in that component, $|(1, 1, \ldots, 1)| = \text{lcm}(n_1, n_2, \ldots, n_t)$ and the order of every element in the product must divide $\text{lcm}(n_1, n_2, \ldots, n_t)$.

3. The mapping $\phi(g) = (g, e_H)$ is an isomorphism from G to $G \oplus \{e_H\}$. To verify that ϕ is one-to-one, we note that $\phi(g) = \phi(g')$ implies $(g, e_H) = (g', e_H)$ which means that $g = g'$. The element $(g, e_H) \in G \oplus \{e_H\}$ is the image of g. Finally, $\phi((g, e_H)(g', e_H)) = \phi((gg', e_He_H)) = \phi((gg', e_H)) = gg' = \phi((g, e_H))\phi((g', e_H))$. A similar argument shows that $\phi(h) = (e_G, h)$ is an isomorphism from H onto $\{e_G\} \oplus H$.

4. $(g, h)(g', h') = (g', h')(g, h)$ for all g, g', h, h' if and only if $gg' = g'g$ and $hh' = h'h$, that is, if and only if G and H are Abelian. A corresponding statement holds for the external direct product of any number of groups.

5. If $Z \oplus Z = \langle(a, b)\rangle$ then neither a nor b is 0. But then $(1, 0) \notin \langle(a, b)\rangle$. $Z \oplus G$ is not cyclic when $|G| > 1$.

6. $Z_8 \oplus Z_2$ contains elements of order 8, while $Z_4 \oplus Z_4$ does not.

7. Define a mapping from $G_1 \oplus G_2$ to $G_2 \oplus G_1$ by $\phi(g_1, g_2) = (g_2, g_1)$. To verify that ϕ is one-to-one, we note that $\phi((g_1, g_2)) = \phi((g_1', g_2'))$ implies $(g_2, g_1) = (g_2', g_1')$. From this we obtain that $g_1 = g_1'$ and $g_2 = g_2'$. The element (g_2, g_1) is the image on (g_1, g_2) so ϕ is onto. Finally, $\phi((g_1, g_2)(g_1', g_2')) = \phi((g_1g_1', g_2g_2')) = (g_2g_2', g_1g_1') = (g_2, g_1)(g_2', g_1') = \phi((g_1, g_2))\phi((g_1', g_2'))$. In general, the external direct product of any number of groups is isomorphic to the external direct product of any rearrangement of those groups.

8. No, $Z_3 \oplus Z_9$ does not have an element of order 27. See also Theorem 8.2.

9. In $Z_{pm} \oplus Z_p$ take $\langle(1, 0)\rangle$ and $\langle(1, 1)\rangle$.

10. Z_9 has 6 elements of order 9 (the members of $U(9)$). Any of these together with any element of Z_3 gives an ordered pair whose order is 9. So $Z_3 \oplus Z_9$ has 18 elements of order 9.

11. In both $Z_4 \oplus Z_4$ and $Z_{8000000} \oplus Z_{400000}$, $|(a, b)| = 4$ if and only if $|a| = 4$ and $|b| = 1, 2$ or 4 or if $|b| = 4$ and $|a| = 1$ or 2 (we have already counted the case that $|a| = 4$). For the first case, we have $\phi(4) = 2$ choices for a and $\phi(4) = \phi(2) + \phi(1) = 4$ choices for b to give us 8 in all. For the second case, we have $\phi(4) = 2$ choices for b and $\phi(2) + \phi(1) = 2$ choices for b. This gives us a total of 12.

    In the general case, observe that by Theorem 4.4 as long as d divides n, the number of elements of order d in a cyclic group depends only on d.

12. $U_5(35) = \{1, 6, 11, 16, 26, 31\}$; $U_7(35) = \{1, 8, 22, 29\}$.

13. Z_{n^2} and $Z_n \oplus Z_n$.

14. The group of rotations is Abelian and a group of order 2 is Abelian; now use Exercise 4.

15. Define a mapping ϕ from $\mathbf{C}$ to $\mathbf{R} \oplus \mathbf{R}$ by $\phi(a + bi) = (a, b)$. To verify that ϕ is one-to-one, note that $\phi(a + bi) = \phi(a' + b'i)$ implies that $(a, b) = (a', b')$. So, $a = a'$ and $b = b'$ and therefore $a + bi = a' + b'i$. The element (a, b) in $\mathbf{R} \oplus \mathbf{R}$ is the image of $a + bi$ so ϕ is onto. Finally, $\phi((a + bi) + (a' + b'i)) = \phi((a + a') + (b + b')i) = (a + a', b + b') = (a, b) + (a', b') = \phi(a + bi) + \phi(a' + b'i)$.

16. Let $\alpha : G_1 \to G_2$ and $\beta : H_1 \to H_2$ be isomorphisms. Then $\gamma : G_1 \oplus H_1 \to G_2 \oplus H_2$ given by $\gamma((g_1, h_1)) = (\alpha(g_1), \beta(h_1))$ is an isomorphism. A corresponding statement holds for the external direct product of any number of groups.

17. By Exercise 3 in this chapter, G is isomorphic to $G \oplus \{e_H\}$ and H is isomorphic to $\{e_G\} \oplus H$. Since subgroups of cyclic groups are cyclic, we know that $G \oplus \{e_H\}$ and $\{e_G\} \oplus H$ are cyclic. In general, if the external direct product of any number of groups is cyclic, each of the factors is cyclic.

18. $\langle(10, 10\rangle$; $\langle 20\rangle \oplus \langle 5\rangle$.

19. $\langle m/r\rangle \oplus \langle n/s\rangle$.

20. Observe that $Z_9 \oplus Z_4 \approx Z_4 \oplus Z_9 \approx \langle 3\rangle \oplus \langle 2\rangle$.

21. Since $\langle(g, h)\rangle \subseteq \langle g\rangle \oplus \langle h\rangle$, a necessary and sufficient condition for equality is that $\text{lcm}(|g|, |h|) = |(g, h)| = |\langle g\rangle \oplus \langle h\rangle| = |g||h|$. This is equivalent to $\gcd(|g|, |h|) = 1$.

22. 48; 6

23. In the general case there are $(3^n - 1)/2$.

24. In this case, observe that $|(a, b)| = 2$ if and only if $|a| = 1$ or 2 and $|b| = 1$ or 2 but not both $|a| = 1$ and $|b| = 1$. So, there are $(m + 2)(n + 1) - 1 = mn + m + 2n + 1$ elements of order 2. For the second part, observe that $|(a, b)| = 4$ if and only if $|a| = 4$ and $|b| = 1$ or 2. So, there are $2(n + 1)$ elements of order 4.

25. Define a mapping ϕ from M to N by $\phi\left(\begin{bmatrix} a & b \\ c & d \end{bmatrix}\right) = (a, b, c, d)$. To verify that ϕ is one-to-one we note that $\phi\left(\begin{bmatrix} a & b \\ c & d \end{bmatrix}\right) = \phi\left(\begin{bmatrix} a' & b' \\ c' & d' \end{bmatrix}\right)$ implies $(a, b, c, d) = (a', b', c', d')$. Thus $a = a', b = b', c = c'$, and $d = d'$. This proves that ϕ is one-to-one. The element (a, b, c, d) is the image of $\begin{bmatrix} a & b \\ c & d \end{bmatrix}$ so ϕ is onto. Finally,
$\phi\left(\begin{bmatrix} a & b \\ c & d \end{bmatrix} + \begin{bmatrix} a' & b' \\ c' & d' \end{bmatrix}\right) = \phi\left(\begin{bmatrix} a + a' & b + b' \\ c + c' & d + d' \end{bmatrix}\right) = (a+a', b+b', c+c', d+d') = (a, b, c, d) + (a', b', c', d') = \phi\left(\begin{bmatrix} a & b \\ c & d \end{bmatrix}\right) + \phi\left(\begin{bmatrix} a' & b' \\ c' & d' \end{bmatrix}\right)$.

Let $\mathbf{R}^k$ denote $\mathbf{R} \oplus \mathbf{R} \oplus \cdots \oplus \mathbf{R}$ (k factors). Then the group of $m \times n$ matrices under addition is isomorphic to R^{mn}.

26. D_6. Since $S_3 \oplus Z_2$ is non-Abelian, it must be isomorphic to A_4 or D_6. But $S_3 \oplus Z_2$ contains an element of order 6 and A_4 does not.

27. Since $(g,g)(h,h)^{-1} = (gh^{-1}, gh^{-1})$, H is a subgroup. When $G = \mathbf{R}$, $G \oplus G$ is the plane and H is the line $y = x$.

28. $D_{12}, S_4, A_4 \oplus Z_2, D_4 \oplus Z_3, D_3 \oplus Z_4, D_3 \oplus Z_2 \oplus Z_2$.

29. $\langle(3,0)\rangle$, $\langle(3,1)\rangle$, $\langle(3,2)\rangle$, $\langle(0,1)\rangle$

30. $\langle(1,0)\rangle$, $\langle(1,1)\rangle$, $\langle(1,2).\rangle$, $\langle(1,3)\rangle$, $\langle(0,1)\rangle$, $\langle(2,1)\rangle$, $\{(0,0),(2,0),(0,2),(2,2)\}$

31. lcm(6,10,15) = 30; $\text{lcm}(n_1, n_2, \ldots, n_k)$.

32. In general, if m and n are even, then $Z_m \oplus Z_n$ has exactly 3 elements of order 2. For if $|(a,b)| = 2$, then $|a| = 1$ or 2 and $|b| = 1$ or 2 but not both, a and b have order 1. Since any cycle group of even order has exactly 1 element of order 2 and 1 of order 1, there are only 3 choices for (a,b).

33. Noting that $Z_4 \oplus Z_3 \oplus Z_2 \approx Z_4 \oplus Z_6$ we find $\langle 15\rangle \oplus \langle 10\rangle$. Noting that $Z_4 \oplus Z_3 \oplus Z_2 \approx Z_2 \oplus Z_{12}$ we find $\langle 50\rangle \oplus \langle 5\rangle$.

34. $\langle 25\rangle \oplus \langle R_{90}\rangle$.

35. Let F be a reflection in D_3. $\{R_0, F\} \oplus \{R_0, R_{180}, H, V\}$.

36. $\langle 4\rangle \oplus \langle 0\rangle \oplus \langle 5\rangle$

37. In $\mathbf{R}^* \oplus \mathbf{R}^*$ $(1,-1)$, $(-1,1)$ and $(-1,-1)$ have order 2, whereas in $\mathbf{C}^*$ the only element of order 2 is -1. But isomorphisms preserve order.

38. $Z_3 \oplus Z_3$.

39. Define the mapping from G to $Z \oplus Z$ by $\phi(3^m6^n) = (m,n)$. To verify that ϕ is one-to-one note that $\phi(3^m6^n) = \phi(3^s6^t)$ implies that $(m,n) = (s,t)$, which in turn implies that $m = s$ and $n = t$. So, $3^m6^n = 3^s6^t$. The element (m,n) is the image of 3^m6^n so ϕ is onto. Finally,
$\phi((3^m6^n)(3^s6^t)) = \phi(3^{m+s}6^{n+t}) = (m+s, n+t) = (m,n) + (s,t) = \phi(3^m6^n)\phi(3^s6^t)$
shows that ϕ is operation preserving.
When $G = \{3^m9^n \mid m,n \in Z\}$ the correspondence from G to $Z \oplus Z$ given by $\phi(3^m9^n) = (m,n)$ is not well-defined since $\phi(3^29^0) \neq \phi(3^09^1)$ and $3^29^0 = 9 = 3^09^1$.

40. $|a_i| = \infty$ for some i.

41. Both D_6 and $D_3 \oplus Z_2$ have 1 element of order 1, 7 of order 2, 2 of order 3, and 2 of order 6. (In fact, they are isomorphic as we see in Example 19 in Chapter 9.)

42. Observe that $U(40) \oplus Z_6 \approx U(8) \oplus U(5) \oplus Z_6 \approx Z_2 \oplus Z_2 \oplus Z_4 \oplus Z_6$ and $U(72) \oplus Z_4 \approx U(9) \oplus U(8) \oplus Z_4 \approx Z_6 \oplus Z_2 \oplus Z_2 \oplus Z_4$ so they are isomorphic.

43. C^* has only one element of order 2 whereas $Z_m \oplus Z_n$ has exactly one element of order 2 if and only if it is cyclic, which is true if and only if $\gcd(m,n) = 1$.

44. If exactly one n_i is even, then x is the unique element of order 2. Otherwise x is the identity.
To see this observe that for each a with $|a| > 2$, a and a^{-1} are distinct, so aa^{-1} contribute e to the product. Let $H = \{a \in G \mid a^2 = e\}$ and let $|H| = 2^n$. Then H is isomorphic to a group of the form $Z_2 \oplus Z_2 \oplus \cdots \oplus Z_2$. The problem is now reduced to H. If $n = 0$ or 1, we are done. If $n \geq 2$, arrange the elements of H without the parentheses and commas in a vertical array with any element at the top, any unused

element below it, any unused element below the second one, and so on. Now let $x = (a_1, a_2, \ldots, a_n)$ be the product of all the elements of H and let $i \in \{1, 2, \ldots, n\}$. Because the set of elements of G that have 0 is the ith column form a subgroup of order 2^{n-1} and the set of elements of G that have 1 the ith column consist of the other 2^{n-1} elements of the column, the sum of all the elements in the ith column is 0. So, $a_i = 0$.

45. Each cyclic subgroup of order 6 has two elements of order 6. So, the 24 elements of order 6 yield 12 cyclic subgroups of order 6. In general, if a group has $2n$ elements of order 6, it has n cyclic subgroups of order 6. (Recall from the Corollary of Theorem 4.4, if a group has a finite number of elements of order 6, the number is divisible by $\phi(6) = 2$).

46. $Z \oplus D_4 \oplus A_4$.

47. $\text{Aut}(U(25)) \approx \text{Aut}(Z_{20}) \approx U(20) \approx U(4) \oplus U(5) \approx Z_2 \oplus Z_4$.

48. S_3

49. In each position we must have an element of order 1 or 2 except for the case that every position has the identity. So, there are $2^k - 1$ choices. For the second question, we must use the identity in every position for which the order of the group is odd. So, there are $2^t - 1$ elements of order 2 where t is the number of $n_1, n_2, \ldots, n_k$ that are even.

50. $Z_{10} \oplus Z_{12} \oplus Z_6 \approx Z_2 \oplus Z_5 \oplus Z_{12} \oplus Z_6 \approx Z_2 \oplus Z_{60} \oplus Z_6 \approx Z_{60} \oplus Z_6 \oplus Z_2$. $Z_{10} \oplus Z_{12} \oplus Z_6$ has 7 elements of order 2 whereas $Z_{15} \oplus Z_4 \oplus Z_{12}$ has only 3.

51. Part **a.** $\phi(18) = 6$; to find an isomorphism all we need do is take 1 to a generator of $Z_2 \oplus Z_9$. So, $\phi(1) = (1, 1)$, which results in $\phi(x) = (x, x)$ Another is $\phi(1) = (1, 2)$, which results in $\phi(x) = (x, 2x)$. Part **b** 0, because $Z_2 \oplus Z_3 \oplus Z_3$ is not cyclic.

52. Since $\phi((2, 3)) = 2$ we have $8\phi((2, 3)) = 16 = 1$. So, $1 = \phi((16, 24)) = \phi((1, 4))$.

53. Since $(2, 0)$ has order 2, it must map to an element in Z_{12} of order 2. The only such element in Z_{12} is 6. The isomorphism defined by $(1, 1)x \to 5x$ with $x = 6$ takes $(2, 0)$ to 6. Since $(1, 0)$ has order 4, it must map to an element in Z_{12} of order 4. The only such elements in Z_{12} are 3 and 9. The first case occurs for the isomorphism defined by $(1, 1)x \to 7x$ with $x = 9$ (recall $(1, 1)$ is a generator of $Z_4 \oplus Z_3$); the second case occurs for the isomorphism defined by $(1, 1)x \to 5x$ with $x = 9$.

54. $U_4(140) \approx U(35) \approx U(5) \oplus U(7) \approx Z_4 \oplus Z_6$.

55. Since $a \in Z_m$ and $b \in Z_n$, we know that $|a|$ divides m and $|b|$ divides n. So, $|(a, b)| = \text{lcm}(|a|, |b|)$ divides $\text{lcm}(m, n)$.

56. Map $ax^2 + bx + c$ to (a, b, c). In general, $\{a_{n-1}x^{n-1} + \cdots + a_0 | \ a_{n-1}, \ldots, a_0 \in Z_m\}$ under addition modulo m is isomorphic to $Z_m \oplus \cdots \oplus Z_m$ (n copies).

57. Up to isomorphism, Z is the only infinite cyclic group and it has 1 and -1 as its only generators. The number of generators of Z_m is $|U(m)|$ so we must determine those m such that $|U(m)| = 2$. First consider the case where $m = p^n$, where p is a prime. Then the number of generators is $p^{n-1}(p - 1)$. So, if $p > 3$, we will have more than 2 generators. When $p = 3$ we must have $n = 1$. Finally, $|U(2^n)| = 2^{n-1} = 2$ only when $n = 2$. This gives us Z_3 and Z_4. When $m = p_1p_2 \cdots p_k$, where the p's are distinct primes, we have $|U(m)| = |U(p_1)||U(p_2)| \cdots |U(p_k)|$. As before, no prime can be greater than 3. So, the only case is $m = 2 \cdot 3 = 6$.

58. Identify A with (0,0), T with (1,1), G with (1,0) and C with (0,1). Then a string of length n of the four bases is represented by a string of 0s and 1s of length $2n$ and the complementary string of $a_1a_2 \ldots a_{2n}$ is $a_1a_2 \ldots a_{2n} + 11 \ldots 1$.

59. Each subgroup of order p consists of the identity and $p-1$ elements of order p. So, we count the number of elements of order p and divide by $p-1$. In $Z_p \oplus Z_p$ every nonidentity element has order p, so there are $(p^2-1)/(p-1) = p+1$ subgroups of order p.

60. $Z \oplus D_3$.

61. In $Z \oplus Z_2$ $|(1,1)| = \infty, |(-1,0)| = \infty, |(1,1)(-1,0)| = |(0,1)| = 2$.

62. $U(165) \approx U(11) \oplus U(3) \oplus U(5) \approx Z_{10} \oplus Z_2 \oplus Z_4$.

63. $U(165) \approx U(15) \oplus U(11) \approx U(5) \oplus U(33) \approx U(3) \oplus U(55) \approx U(3) \oplus U(5) \oplus U(11)$.

64. Note that $U_9(72) \approx U(8) \approx Z_2 \oplus Z_2$ and
$U_4(300)) \approx U(75) \approx U(3) \oplus U(25) \approx Z_2 \oplus Z_{20}$.

65. From Theorem 8.3 and Exercise 3 we have $U(2n) \approx U(2) \oplus U(n) \approx U(n)$.

66. Since $U(2^n)$ is isomorphic to $Z_{2^{n-2}} \oplus Z_2$, and $Z_{2^{n-2}}$ and Z_2 each have exactly one element of order 2, $U(2^n)$ has exactly three elements of order 2.

67. We use the fact that $\mathrm{Aut}(Z_{105}) \approx U(105) \approx U(3) \oplus U(5) \oplus U(7) \approx Z_2 \oplus Z_4 \oplus Z_6$. In order for (a,b,c) to have order 6, we could have $|c| = 6$ and a and b have orders 1 or 2. So we have 2 choices for each of a, b, and c. This gives 8 elements. The only other possibility for (a,b,c) to have order 6 is for $|c| = 3$ and a and b have orders 1 or 2, but not both have order 1. So we have 3 choices for a and b together and 2 choices for c. This gives 6 more elements for a total of 14 in all. For the second part, use the fact that $U(27) \approx Z_{18}$.

68. By Theorem 6.5 we have
$\mathrm{Aut}(\mathrm{Aut}(Z_{50})) \approx \mathrm{Aut}(U(50)) \approx \mathrm{Aut}(U(50)) \approx \mathrm{Aut}(Z_{20}) \approx U(20) \approx Z_2 \oplus Z_4$.

69. $A_4 \oplus Z_4$ has 7 elements of order 2 whereas the subgroup $D_{12} \oplus \{0\}$ of $D_{12} \oplus Z_2$ has 13.
Alternate solution 1. $A_4 \oplus Z_4$ has 16 elements of order 12 whereas $D_{12} \oplus Z_2$ has 8. (Note that these are consistent with the corollary of Theorem 4.4.)
Alternate solution 2. By Exercise 77 in Chapter 5 $|Z(A_4 \oplus Z_4)| = 1$, whereas $(R_{180}, 0)$ is in the center of $D_{12} \oplus Z_2$.

70. $D_8 \oplus D_3$ has elements of order 8 and $D_6 \oplus D_4$ does not.

71. $Z \oplus D_4$

72. $U(p^m) \oplus U(q^n) = Z_{p^m - p^{m-1}} \oplus Z_{q^n - q^{n-1}}$ and both of these groups have even order. Now use Theorem 8.2.

73. Observe that $U(55) \approx U(5) \oplus U(11) \approx Z_4 \oplus Z_{10}$ and
$U(75) \approx U(3) \oplus U(25) \approx Z_2 \oplus Z_{20} \approx Z_2 \oplus Z_5 \oplus Z_4 \approx Z_{10} \oplus Z_4 \approx Z_4 \oplus Z_{10}$.
$U(144) \approx U(16) \oplus U(9) \approx Z_4 \oplus Z_2 \oplus Z_6$;
$U(140) \approx U(4) \oplus U(5) \oplus U(7) \approx Z_2 \oplus Z_4 \oplus Z_6$.

74. $U(900) \approx Z_2 \oplus Z_6 \oplus Z_{20}$, so the element of largest order is the lcm(2, 6, 20) = 60.

75. From Theorem 6.5 we know $\text{Aut}(Z_n) \approx U(n)$. So, $n = 8$ and 12 are the two smallest.

76. Observe that $Z_2 \oplus Z_4 \oplus Z_9 \approx Z_4 \oplus Z_2 \oplus Z_9 \approx Z_4 \oplus Z_{18} \approx U(5) \oplus U(27) \approx U(135)$.

77. Since $U(pq) \approx U(p) \oplus U(q) \approx Z_{p-1} \oplus Z_{q-1}$ if follows that $k = \text{lcm}(p-1, q-1)$.

78. $U_{50}(200) = \{1, 51, 101, 151\}$ has order 4 whereas $U(4)$ has order 2. $|U(200)| = 80$; $|U(50) \oplus U(4)| = 40$. This is not a contradiction to Theorem 8.3 because 50 and 4 are not relatively prime.

79. $Z_{p^{n-1}}$. To see this, first observe that because $U(p^n)$ is cyclic, so is $U_p(p^n)$. Now list the elements of $U_p(p^n)$ as follows: $1+p, 1+2p, \ldots, 1+p^{n-1}p = 1+p^n = 1$. This gives us p^{n-1} elements.

80. Since $U(p^k)$ is cyclic and $|U(p^k)| = p^k - p^{k-1} = p^{k-1}(p-1)$, we know from Theorem 4.3 and Lagrange's theorem that $U(p^k)$ has an element of order if and only if 4 divides $p-1$.

81. $U_8(40) \approx U(5) \approx Z_4$.

82. In the first case, there are $2^k - 1$; in the second case, there are $2^{k+2} - 1$.

83. $U_5(140) \approx U(28)$; $U_4(140) \approx U(35) \approx Z_4 \oplus Z_6$.

84. 3, 6, 8, 12, 24.

85. If $x \in U_{st}(n)$, then $x \in U(n)$ and $x - 1 = stm$ for some m. So, $x - 1 = s(tm)$ and $x - 1 = t(sm)$. If $x \in U_s(n) \cap U_t(n)$, then $x \in U(n)$ and both s and t divide $x - 1$. So, by Exercise 6 in Chapter 0, st divides $x - 1$.

86. This follows directly from Cayley's Theorem (6.5 in Chapter 6).

87. $Z_2 \oplus Z_2$.

88. $Z_2 \oplus Z_2 \oplus Z_2$.

89. Let $p_1^{n_1} p_2^{n_2} \cdots p_k^{n_k}$ be the prime-power decomposition of n. From the discussion following the corollary of Theorem 9.7, we know that $U(n)$ is the internal direct product of groups $U(p_i^{n_i})$ for $i = 1, 2, \ldots, k$. For odd p_i, $U(p_i^{n_i})$ is cyclic of even order so H contains the generator the subgroup isomorphic to $U(p_i^{n_i})$. For $p_i = 2$, every nonidentity element of $U(p_i^{n_i})$ has even order so H contains $U(2^{n_i})$ as well.

90. First observe that $U(n)^k = U(n)$ if and only if the mapping $x \to x^k$ is $1-1$. Let $d = \gcd(|U(n|), k)$. If $d > 1$, let p be a prime divisor of d. By Theorem 9.5, $U(n)$ has an element a of order p. Then $a^p = 1$ and therefore $a^k = 1$. Now suppose that $d = 1$. If $a, b \in U(n)$ and $a^k = b^k$, we have $(ab^{-1})^k = 1$. Thus, $|ab^{-1}|$ is a common divisor of $|U(n)|$ and k. This forces $ab^{-1} = 1$ and $a = b$.

91. Since $5 \cdot 29 = 1 \bmod 36$, we have that $s = 29$. So, we have $34^{29} \bmod 2701 = 1415$, which converts to NO.

92. None. Because gcd(18,12) = 6, Step 3 of the Sender part of the algorithm fails.

93. Because the block 2505 exceeds the modulus 2263, sending $2505^e \bmod 2263$ is the same as sending $242^e \bmod 2263$, which decodes as 242 instead of 2505.

CHAPTER 9

Normal Subgroups and Factor Groups

1. No, $(13)(12)(13)^{-1} = (23)$ is not in H.
2. For every α in S_n, $\alpha A_n \alpha^{-1}$ is even.
3. $HR_{90} = R_{270}H; DR_{270} = R_{90}D; R_{90}V = VR_{270}$
4. Solving $(12)(13)(14) = \alpha(12)$ for α we have $\alpha = (12)(13)(14)(12)$. Solving $(1234)(12)(23) = \alpha(1234)$ for α we have $\alpha = (234)$.
5. Say $i < j$ and $h \in H_i \cap H_j$. Then $h \in H_1H_2 \cdots H_{j-1} \cap H_j = \{e\}$.
6. No. Let $A = \begin{bmatrix} 1 & 0 \\ 0 & -1 \end{bmatrix}$ and $B = \begin{bmatrix} 1 & 0 \\ 1 & 1 \end{bmatrix}$. Then A is in H and B is in $GL(2, \mathbf{R})$ but BAB^{-1} is not in H.
7. H contains the identity so H is not empty. Let $A, B \in H$. Then det $(AB^{-1}) =$ (det A)(det $B)^{-1} \in K$. This proves that H is a subgroup. Also, for $A \in H$ and $B \in G$ we have det $(BAB^{-1}) =$ (det B)(det A)(det $B)^{-1} =$ det $A \in K$ so $BAB^{-1} \in H$.
8. If k divides n, then $\langle k \rangle / \langle n \rangle$ is a cyclic group of order n/k. So, it is isomorphic to $Z_{n/k}$.
9. Let $x \in G$. If $x \in H$, then $xH = H = Hx$. If $x \notin H$, then xH is the set of elements in G, not in H and Hx is also the elements in G, not in H.
10. **a.** $\alpha_5 H \neq H\alpha_5$. **b.** It proves that coset multiplication is not a binary operation.
11. Let $G = \langle a \rangle$. Then $G/H = \langle aH \rangle$.
12. Note that $aHbH = abH = baH = bHaH$.
13. in H.
14. 4
15. $|9H| = 2$; $|13H| = 4$.
16. 3
17. $8 + \langle 3.5 \rangle = 1 + 7 + \langle 3.5 \rangle = 1 + \langle 3.5 \rangle$.
18. Z_k
19. Observe that in a group G, if $|a| = 2$ and $\{e, a\}$ is a normal subgroup, then $xax^{-1} = a$ for all x in G. Thus $a \in Z(G)$. So, the only normal subgroup of order 2 in D_n is $\{R_0, R_{180}\}$ when n is even.
20. Since $|U(40)| = 16$ and $U_5(40) = \{1, 11, 21, 31\}$ and $U_8(40) = \{1, 9, 17, 33\}$ we know that both factor groups have order 4. In $U_5(40)$, $3U_8(40)$ has order 4, whereas all three non-identity elements in $U(40)/U_8(40)$ have order 2.

21. Observe that (57) is in H, whereas (56)(57)(56) sends 6 to 7. Observe that (578) is in H, whereas (456)(57)(654) sends 6 to 7.

22. Note that
$(x, y) + \langle(2,2)\rangle = (x-y, 0) + (y, y) + \langle(2,2)\rangle = (x-y)(1,0) + y(1,1) + \langle(2,2)\rangle$. If y is even, then $(y, y) + \langle(2,2)\rangle = \langle(2,2)\rangle$. If y is odd, then
$(y, y) + \langle(2,2)\rangle = (1,1) + \langle(2,2)\rangle$ $(Z \oplus Z)/\langle(2,2)\rangle \approx Z \oplus Z_2$;
$(Z \oplus Z)/\langle(n,n)\rangle \approx Z \oplus Z_n$.

23. It follows from Theorem 7.2 that H is the unique subgroup of order p in G. Then, since for every x in G, we have that xHx^{-1} is a subgroup of order p, $xHx^{-1} = H$.

24. $Z_4 \oplus Z_2$. To see that there is no element of order 8 in the factor group, observe that for any element (a, b) in $Z_4 \oplus Z_{12}$, $(a, b)^4 = (4a, 4b)$ belongs to $\{(0,0), (0,4), (0,8)\} \in \langle(2,2)\rangle$. So, the order of every element in the factor group divides 4. This rules out Z_8. By observation, $(1,0)\langle(2,2)\rangle$ has order 4, which rules out $Z_2 \oplus Z_2 \oplus Z_2$.

25. Since the element $(3H)^4 = 17H \neq H$, $|3H| = 8$. Thus $G/H \approx Z_8$.

26. $Z_4 \oplus Z_2$; $\langle 17\rangle \times \langle 41\rangle$.

27. Since H and K have order 2, they are both isomorphic to Z_2 and therefore isomorphic to each other. Since $|G/H| = 4$ and $|3H| = 4$ we know that $G/H \approx Z_4$. On the other hand, direct calculations show that each of the three nonidentity elements in G/K has order 2, so $G/K \approx Z_2 \oplus Z_2$.

28. $Z_2 \oplus Z_2$; Z_4.

29. Observe that nontrivial proper subgroups of a group of order 8 have order 2 or 4 and therefore are Abelian. Then use Theorem 9.6 and Exercise 4 of Chapter 8.

30. $U(165) = U_{15}(165) \times U_{11}(165) = U_{33}(165) \times U_5(165) = U_{55}(165) \times U_3(165)$.

31. Certainly, every nonzero real number is of the form $\pm r$, where r is a positive real number. Real numbers commute and $\mathbf{R}^+ \cap \{1, -1\} = \{1\}$.

32. By Corollary 4 of Theorem 7.1, $x^m N = (xN)^m = N$, so $x^m \in N$.

33. In the general case that $G = HK$ there is no relationship. If $G = H \times K$, then $|g| = \text{lcm}(|h|, |k|)$, provided the $|h|$ and $|k|$ are finite. If $|h|$ or $|k|$ is infinite, so is $|g|$.

34. Since $1 = 3 \cdot 5 + (-2)7$, $m = (3m)5 + (-2m)7$. No, since $35 \in H \cap K$. Suppose H is any subgroup of index 2. Then, since $|\mathbf{R}^*/H| = 2$, we have $a^2 \in H$ for all $a \in \mathbf{R}^*$. This implies that $\mathbf{R}^+ \subseteq H$. If there is some $a \in H$ with $a < 0$, then since $-a \in H$ we have $a^{-1}(-a) = -1 \in H$ also. But this implies that $\mathbf{R}^* \subseteq H$.

35. For the first question, note that $\langle 3\rangle \cap \langle 6\rangle = \{1\}$ and $\langle 3\rangle\langle 6\rangle \cap \langle 10\rangle = \{1\}$. For the second question, observe that $12 = 3^{-1}6^2$ so $\langle 3\rangle\langle 6\rangle \cap \langle 12\rangle \neq \{1\}$.

36. Certainly, $\mathbf{R}^+$ has index 2. Suppose that H has index 2 and is not $\mathbf{R}^+$. Then $|\mathbf{R}^*/H| = 2$. So, for every nonzero real number a we have $(aH)^2 = a^2H = H$. Thus the square of every real number is in H. This implies that H contains all positive real numbers. Since H is not $\mathbf{R}^+$, it must contain some negative real number a. But then H contains $a\mathbf{R}^+$, which is the set of all negative real numbers. This shows that $H = \mathbf{R}^*$.

37. Say $|g| = n$. Then $(gH)^n = g^nH = eH = H$. Now use Corollary 2 to Theorem 4.1.

38. Observe that $(1/n + Z)$ has order n.

39. Let x belong to G. Then $xHx^{-1}H = xx^{-1}H = H$ so $xHx^{-1} \subseteq H$.
Alternate solution: Let x belong to G and h belong to H. Then
$xhx^{-1}H = xhHx^{-1})H = xHx^{-1}H = xx^{-1}H = eH = H$, so xhx^{-1} belongs to H.

40. Observe that $(13)H(23)H = \{(132), (23), (12)\}$ whereas $(13)(23)H = \{(13), (123)\}$.

41. Suppose that H is a proper subgroup of Q of index n. Then Q/H is a finite group of order n. By Corollary 4 of Theorem 7.1 we know that for every x in Q we have nx is in H. Now observe that the function $f(x) = nx$ maps Q onto Q. So, $Q \subseteq H$.

42. Let $g \in G$. Then there is an element b in G so that $gH = (bH)^2 = b^2H$. Thus $g = b^2h$ for some $h \in H$. But there is a c in H such that $h = c^2$ So, $g = (bc)^2$. The proof is valid for any integer.

43. Take $G = Z_6$, $H = \{0, 3\}$, $a = 1$, and $b = 4$.

44. That ϕ is 1-1 and onto is obvious. Moreover,
$\phi((h_1h_2\cdots h_n)(h_1'h_2'\cdots h_n')) = \phi(h_1h_1'h_2h_2'\cdots h_nh_n') = (h_1h_1', h_2h_2', \ldots, h_nh_n') =$
$(h_1, h_2, \ldots, h_n)(h_1', h_2', \ldots, h_n') = \phi((h_1h_2\cdots h_n)\phi(h_1'h_2'\cdots h_n')$.

45. Normality follows directly from Theorem 4.3 and Example 7.

46. By Example 14 of Chapter 3, $Z(D_{13})$ is the identity. Then by Theorem 9.4, D_{13} is isomorphic to $\text{Inn}(D_{13})$.

47. By Lagrange, $|H \cap K|$ divides both 63 and 45. If $|H \cap K| = 9$, then $H \cap K$ is Abelian by Theorem 9.7. If $|H \cap K| = 3$, then $H \cap K$ is cyclic by the Corollary of Theorem 7.1. $|H \cap K| = 1$, then $H \cap K = \{e\}$. In general, if p is a prime and $|H| = p^2m$ and $|K| = p^2n$ where $\gcd(m, n) = 1$, then $H \cap K| = 1, p$, or p^2. So by Corollary 3 of Theorem 7.1 and Theorem 9.7, $H \cap K$ is Abelian.

48. Use Example 3 in Chapter 8 and the G/Z Theorem (Theorem 9.3).

49. By Lagrange's Theorem, $|Z(G)| = 1, p, p^2$, or p^3. By assumption, $|Z(G)| \neq 1$ or p^3 (for then G would be Abelian). So, $|Z(G)| = p$ or p^2. However, the "G/Z" Theorem (Theorem 9.3) rules out the latter case.

50. Use the G/Z Theorem.

51. Suppose that K is a normal subgroup of G and let $gH \in G/H$ and $kH \in K/H$. Then $gHkH(gH)^{-1} = gHkHg^{-1}H = gkg^{-1}H \in K/H$. Now suppose that K/H is a normal subgroup of G/H and let $g \in G$ and $k \in K$. Then
$gkg^{-1}H = gHkHg^{-1}H = gHkH(gH)^{-1} \in K/H$ so $gkg^{-1} \in K$.

52. Say $|aH|$ has finite order n. Then $H = (aH)^n = a^nH$ so that a^n is in H. But this implies that $|a^n|$ and therefore $|a|$ is finite. Thus $aH = H$.

53. Say H has an index n. Then $(\mathbf{R}^*)^n = \{x^n \mid x \in \mathbf{R}^*\} \subseteq H$. If n is odd, then $(\mathbf{R}^*)^n = \mathbf{R}^*$; if n is even, then $(\mathbf{R}^*)^n = \mathbf{R}^+$. So, $H = \mathbf{R}^*$ or $H = \mathbf{R}^+$.

54. **a.** Since 1 and -1 commutes with every element of the group, H is normal in G.

b.

	1	i	j	k
1	1	i	j	k
i	i	1	k	j
j	j	k	1	i
k	k	j	i	1

$Z_2 \oplus Z_2$

55. By Exercise 9, we know that K is normal in L and L is normal in D_4. But $VK = \{V, R_{270}\}$, whereas $KV = \{V, R_{90}\}$. So, K is not normal in D_4.

56. $x(H \cap N)x^{-1} = xHx^{-1} \cap xNx^{-1} = H \cap N$. The same argument works for the intersection of any family of normal subgroups.
Alternate proof. Let $a \in H \cap K$ and $x \in G$. Then, because H and K are normal in G, we have $xax^{-1} \in H$ and $xax^{-1} \in K$. So $xax^{-1} \in H \cap K$.

57. G has elements of orders 1, 2, 3, and 6.

58. By Example 5, NM is a subgroup. Also

$$xNM = (xN)M = (Nx)M = N(xM) = N(Mx) = NMx.$$

59. Let $H = \langle a^k \rangle$ be any subgroup of $N = \langle a \rangle$. Let $x \in G$ and let $(a^k)^m \in H$. We must show that $x(a^k)^m x^{-1} \in H$. Note that
$x(a^k)^m x^{-1} = x(a^{km})x^{-1} = (xax^{-1})^{km} = (a^r)^{km} = (a^k)^{rm} \in \langle a^k \rangle$. (Here we used the normality of N to replace xax^{-1} by a^r.)

60. Let H be the unique maximal subgroup of G and suppose that G is not normal in G. Let a be any element in G such that $aHa^{-1} \neq H$. Then aHa^{-1} is not a maximal subgroup of G. Let K be a proper subgroup of G that properly contains aHa^{-1}. Then, $|aHa^{-1}| < |K| < |G|$. So, applying the inner automorophism $\phi_{(a^{-1})}$ induced by a^{-1} to $aHa^{-1} \subset K \subset G$, we have $H \subset a^{-1}Ka \subset G$, which is a contradiction.

61. $\gcd(|x|, |G|/|H|) = 1$ implies $\gcd(|xH|, |G/H|) = 1$. But $|xH|$ divides G/H. Thus $|xH| = 1$ and therefore $xH = H$.

62. It follows from Theorem 7.2 that H is the only subgroup of G of order p and from Example 7 that H is normal.

63. If H and K are subgroups of order 3 and one of them is normal then HK is a subgroup of order 9. This contradicts Lagrange.

64. **a.** Observe that for any g in G,

$$g(x^{-1}y^{-1}xy)g^{-1} = (gx^{-1}g^{-1})(gy^{-1}g^{-1})(gxg^{-1})(gyg^{-1}) \in S.$$

b. Observe that $xG'yG' = yG'xG'$ if and only if $x^{-1}y^{-1}xy \in G'$.
c. Observe that $xNyN = yNxN$ implies $x^{-1}y^{-1}xyN = N$. Thus $x^{-1}y^{-1}xy \in N$.
d. Let $h \in H$ and $g \in G$. Then $ghg^{-1}h^{-1} \in G' \leq H$ so that $ghg^{-1} \in Hh = H$.

65. Observe that for every positive integer n, $(1+i)^n$ is not a real number. So, $(1+i)\mathbf{R}^*$ has infinite order.

66. Let C be the collection of all subgroups of G of order n. Then, since $|H| = n$ implies $|xHx^{-1}| = n$, we have

$$x(\cap_{H \in C} H)x^{-1} = \cap_{H \in C} xHx^{-1} = \cap_{H \in C} H.$$

67. Suppose that Aut(G) is cyclic. Then Inn(G) is also cyclic. So, by Theorem 9.4, G/Z is cyclic and from Theorem 9.3 it follows that G is Abelian. This is a contradiction.

68. Let $x \in K$. Then, $|x| = p^i$ for some i. So $(xH)^{p^i} = x^{p^i}H = H$ and, because $|G/H| = m$, also we have $(xH)^m = H$. Thus $|xH|$ divides both m and p^i. From $\gcd(p, m) = 1$, we have $|xH| = 1$ and therefore $xH = H$. This proves that $K \subseteq H$.

69. Because $|g| = 16$ implies that $|gH|$ divides 16, it suffices to show that $(gH)^4$ is not H. Suppose that $(gH)^4 = g^4H = H$. Then g^4 is in H. But then 2, g^4, and g^8 are in H, which is a contradiction. In the general case, say $|gH| = k$. Then $(gH)^k = g^kH = H$. So, g^k is in H and therefore $|g^k| = 1$ or 2. It follows that $k = 2n$ or n.

70. Suppose that H is a subgroup of S_4 of order 12 distinct from A_4. Then, Example 5 in this chapter and Theorem 7.2 $HA_4 = S_4$ and $|HA_4| = 12 \cdot 12/|H \cap A_4|$. It follows that $|H \cap A_4| = 6$. But this contradicts Example 5 of Chapter 7.

71. If $|Z(G)| = 5$, then $|G/Z(G)| = 6$ and by Theorem 7.3 G is isomorphic to Z_6 or D_6. Theorem 9.3 rules out Z_6. If $|Z(G)| = 3$, then $|G/Z(G)| = 10$ and by Theorem 7.3 G is isomorphic to Z_{10} or D_5. Theorem 9.3 rules out Z_{10}. If $|G| = 2pq$ where p and q are distinct odd primes and $|Z(G)| = p$ or q, then $G/Z(G)$ is isomorphic to D_q or D_p, respectively.

72. By the classification of groups of order p^2-theorem (9.3), G would be cyclic or isomorphic to $Z_p \oplus Z_p$. But the G/Z-Theorem (9.3) rules out the cyclic case.

73. If A_5 had a normal subgroup of order 2, then the subgroup has a nonidentity element that commutes with every element of A_5. An element of A_5 of order 2 has the form $(ab)(cd)$. But $(ab)(cd)$ does not commute with (abc), which also belongs to A_5.

74. Since H has index 2 in G is it a normal subgroup of G and $|G/H| = 2$. It follows that for every a in G we have $(aH)^2 = H$. If a is an element of G of order $2n + 1$, then $H = a^{2n+1}H = ((aH)^2)^n aH = aH$. Thus, a is in H.

75. By the classification of groups of order $2p$-theorem (7.3), $G/Z(G)$ would be cyclic or isomorphic to D_p. But the G/Z-Theorem (9.3) rules out the cyclic case for otherwise G would be Abelian and therefor G would be $Z(G)$.

76. $|9H| = 2$; $|17H| = 4$.

77. Suppose that H is a normal subgroup of A_5 of order 12. Since $|A_5/H| = 5$ we know that for any of the 20 3-cycles α in A_5 we have $H = (\alpha H) = \alpha^5 H = \alpha^2 H$. So, $\alpha^{-1} = \alpha^2 \in H$. Then α is also in H. But H only has 12 elements.

78. Because $U(60) \approx U(4) \oplus U(3) \oplus U(5) \approx Z_2 \oplus Z_2 \oplus Z_4$ every element of $U(60)$ satisfies the equations $x^4 = 1$. So for all $x \in U(60)$ we have would have $(xH)^4 = x^4H = H$.

79. Because $51H = H$ we have $153H = (3 \cdot 51)H = 3(51H) = 3H$.

80. By the $2p$-theorem, $G/Z(G)$ is isomorphic to Z_{14} or D_7. The $G/Z(G)$-theorem rules of Z_{14}. Since D_7 has 7 elements of order 2, so does $G/Z(G)$.

81. Let $|gH| = d$ and $|g| = m$. We know by Exercise 37 that $|gH|$ divides $|g|$ and because $g^d \in H$ we also know that $|g^d| = m/d$ divides $|H|$. This means that $m/d = 1$.

82. Let $H = U_{16}(80) = \{1, 17, 33, 49\}$. Since none of $3, 3^2, 3^3$ is in H and 3^4 is in H the element $3H$ in the factor group $U(80)/H$ has order 4. Thus $\langle 3H \rangle$ is a subgroup of $U(80)/H$ of order 4 and $H \cup 3H \cup 9H \cup 27H$ is a subgroup of $U(80)$ order 16.

83. By definition, every element of G can be written in the form $a_{j_1}a_{j_2}\dots a_{j_k}$ where $a_{j_1}, a_{j_2}, \dots, a_{j_k} \in \langle a_1, a_2, \dots, a_n\rangle$. Then $gH = a_{j_1}Ha_{j_2}H, \dots, a_{j_k}H$.

84. If n is odd then the n rotations of D_n form a normal subgroup of order n (see Exercise 9) that does not contain R_{180}. So, $D_{2n} = D_n \times \{R_0, R_{180}\}$. If n is even, then the elements of D_n contains R_{180} and $\{R_0, R_{180}\}$ is the only normal subgroup of order 2 in D_{2n} (see Exercise 19.)

85. Since G is Abelian, the subgroups $H_1, H_2, \dots, H_k$ are normal in G. By assumption, $G = H_1H_2\cdots H_k$, So, all that remains to prove is that for all $i = 2, 3, \dots, k-1$ we have $H_1H_2\cdots H_i \cap H_{i+1} = \{e\}$. But if $x \in H_1H_2\cdots H_i \cap H_{i+1}$ and $x \neq e$, then x can be written in the two distinct forms $h_1h_2\cdots h_ie\cdots e$ $(k-i$ e terms) and $e\cdots eh_{i+1}e\cdots e$ (i e terms) on the left and $k-(i+1)$ e terms on the right and each $h_j \in H_j$. This contradicts our assumption about G.

86. $eHe^{-1} = H$. Let $x, y \in N(H)$. Then $(xy^{-1})H(xy^{-1})^{-1} = x(y^{-1}Hy)x^{-1} = xHx^{-1} = H$. (Note that $yHy^{-1} = H$ implies $y^{-1}Hy = H$.)

87. We know from Theorem 9.7 that $G/Z(G) \approx Z_{p^2}$ or $Z_p \oplus Z_p$ and from the G/Z Theorem (Theorem 9.3) Z_{p^2} is ruled out.

88. Theorem 9.3 would imply that $G = Z(G)$.

89. By Theorem 7.2 and Example 5 in Chapter 9, if H and K were distinct subgroups of order p^2, then HK would be a subgroup of order p^3 or p^4, which contradicts Lagrange.

90. Let $\alpha \in$ Aut(G) and $\phi_a \in$ Inn(G). Then, $(\alpha\phi_a\alpha^{-1})(x) = (\alpha\phi_a)(\alpha^{-1}(x)) = \alpha(\phi_a(\alpha^{-1}(x)) = \alpha(a\alpha^{-1}(x)a^{-1}) = \alpha(a)x\alpha(a^{-1}) = \alpha(a)x\alpha(a)^{-1}$.

91. If G is cyclic, then Theorem 4.4 says that G has exactly one element of order 2. If G is not cyclic, let a be any non-identity element of G and b be any element of G not in $\langle a\rangle$. Then $\langle a\rangle \times \langle b\rangle$ is isomorphic to a group of the form $Z_{2^s} \oplus Z_{2^t}$ where s and t are positive. But then G has at least three elements of order 2. The appropriate generalization is: "An Abelian group of order p^n for a prime p and some positive integer n is cyclic if and only if it has exactly $p-1$ elements of order p."

92. First note that the argument in Example 4 of Chapter 3 shows that G^m and G^n are subgroups of G and since G is Abelian they are normal in G. Because gcd$(m, n) = 1$, there are integers a and b such that $am + bn = 1$. By closure $G^mG^n \subseteq G$. To show that $G \subseteq G^mG^n$ let $x \in G$. Then we have that $x = x^{am+bn} = x^{am}x^{bn}$, $(x^{am})^n = (x^a)^{mn} = (x^a)^{|G|} = e$, and $(x^{bn})^m = (x^b)^{mn} = (x^b)^{|G|} = e$. So, $x \in G^mG^n$. Finally, if $x \in G^m \cap G^n$ Then $x^m = e = x^n$ and therefore $|x|$ divides both m and n. So, $|x| = 1$.

93. Observe that for every two distinct primes p and q we have $pH \neq qH$. (For if there are integers a, b, c, d such that $pa^2/b^2 = qc^2d^2$, then p occurs an odd number of times on the left side of $pa^2d^2 = qb^2d^2$ but an even number of times on the right side). Every nonidentity element in Q/H has order 2.

CHAPTER 10

Group Homomorphisms

1. Note that det $(AB) = (\det A)(\det B)$.

2. Observe that $|ab| = |a|\,|b|$.

3. Note that $(f+g)' = f' + g'$.

4. Let E denote any even permutation and O any odd. Observe that $\phi(EE) = \phi(E) = 0 = 0+0 = \phi(E)+\phi(E)$. $\phi(EO) = \phi(O) = 1 = 0+1 = \phi(E)+\phi(O)$. The other cases are similar.

5. By property 3 of Theorem 10.1, $|k|$ divides both m and n.
   Now suppose that $|k|$ divides both m and n and let $d = \gcd(m,n)$. Since $|k|$ divides both m and n, it divides d, and because the subgroup $\langle n/d\rangle$ has order d, every element whose order divides d is contained in $\langle n/d\rangle$. So, $k = tn/d$ for some integer t. To show that ϕ is well-defined we must show that if $a = b \mod m$, then $ka = kb \mod n$. Since $a - b = mq$ for some integer q, we have
   $ka - kb = kmt = (tn/d)mt = tn(m/d)t = 0 \mod n$.
   To see why the number of homomorphisms from Z_m to Z_n is $\gcd(m,n)$ observe that the above argument shows that k yields a homomorphism if and only if k is in $\langle n/d\rangle$ and the order of n/d is d.

6. Recall, $\int(f+g) = \int f + \int g$. Kernel $= \{0\}$. No.

7. $(\sigma\phi)(g_1g_2) = \sigma(\phi(g_1g_2)) = \sigma(\phi(g_1)\phi(g_2)) = \sigma(\phi(g_1))\sigma(\phi(g_2)) = (\sigma\phi)(g_1)(\sigma\phi)(g_2)$. It follows from Theorem 10.3 that $|G/\text{Ker }\phi| = |H|$ and $|G/\text{Ker }\sigma\phi| = |K|$. Thus, $[\text{Ker }\sigma\phi : \text{Ker }] = |\text{Ker }\sigma\phi/\text{Ker }\phi| = |H|/|K|$.

8. See Exercise 16 of Chapter 5. The kernel is the set of even permutations in G. When G is S_n the kernel is A_n and from Theorem 10.3 we have that S_n/A_n is isomorphic to $\{+1,-1\}$. So, A_n has index 2 in S_n and is normal in S_n. The kernel is the subgroup of even permutations in G. If the members of G are not all even then the coset other than the kernel is the set of odd permutations in G. All cosets have the same size.

9. $\phi((g,h)(g',h')) = \phi((gg',hh')) = gg' = \phi((g,h))\phi((g',h'))$. The kernel is $\{(e,h) \mid h \in H\}$.

10. See Exercise 9 of Chapter 1. The kernel is the subgroup of rotations in G. If the members of G are not all rotations then the coset other than the kernel is the set of reflections in G. All cosets have the same size.

11. The mapping $\phi : Z \oplus Z \to Z_a \oplus Z_b$ given by $\phi((x,y)) = (x \mod a, y \mod b)$ is operation preserving by Exercise 9 in Chapter 0. If $(x,y) \in \text{Ker }\phi$, then $x \in \langle a\rangle$ and $y \in \langle b\rangle$. So, $(x,y) \in \langle (a,0)\rangle \times \langle (0,b)\rangle$. Conversely, every element in $\langle (a,0)\rangle \times \langle (0,b)\rangle$ is in Ker ϕ. So, by Theorem 10.3, $Z \oplus Z \to Z_a \oplus Z_b$ is isomorphic to $\langle (a,0)\rangle \times \langle (0,b)\rangle$.

12. $x \to x \mod k$ is a homomorphism with kernel $\langle k\rangle$. So, $Z_n/\langle k\rangle$ is a cyclic group of order n/k.

13. $(a, b) \to b$ is a homomorphism from $A \oplus B$ onto B with kernel $A \oplus \{e\}$. So, by Theorem 10.3, $(A \oplus B)/(A \oplus \{e\}) \approx B$.

14. Observe that since 1 has order 12, $|\phi(1)| = |3|$ must divide 12. But in Z_{10}, $|3| = 10$. Alternate proof. Observe that $\phi(6+7) = \phi(1) = 3$ while $\phi(6) + \phi(7) = 8 + 1 = 9$. Second alternate proof. Observe that $\{0, 6\}$ is a subgroup of Z_{12} but $\phi(\{0, 6\}) = \{0, 8\}$ is not a subgroup of Z_{10}.

15. By property 6 of Theorem 10.1, we know $\phi^{-1}(9) = 23 + \text{Ker } \phi = \{23, 3, 13\}$.

16. Observe that such a mapping would be an isomorphism and isomorphisms preserve orders of elements.

17. Suppose ϕ is such a homomorphism. By Theorem 10.3, $|\text{Ker } \phi| = 2$. Let $\phi(1, 0) = (a, b)$. Then $\phi(4, 0) = 4\phi(1, 0) = 4(a, b) = (4a, 4b) = (0, 0)$. But then Ker ϕ contains an element of order 4.
Alternate proof. Suppose ϕ is such a homomorphism and let $H = \text{Ker } \phi$. By Theorem 10.3, $(Z_{16} \oplus Z_2)/H \approx Z_4 \oplus Z_4$. Thus every element of $(Z_{16} \oplus Z_2)/H$ has order 1, 2 or 4 and $|H| = 2$. Then $((1, 0)H)^4 = (4, 0)H = H$ implies that $(4, 0) \in H$. But (4,0) has order 4 whereas $|H| = 2$.

18. No, because of part 3 of Theorem 10.1. No, because the homomorphic image of a cycle group must be cyclic. Yes, $\phi(x) = (x \bmod 3, x \bmod 2)$ is a homomorphism.

19. Since $|\text{Ker } \phi|$ is not 1 and divides 17, ϕ is the trivial map.

20. 0 onto Z_8; 4 to Z_8.

21. By Theorem 10.3 we know that $|Z_{30}/\text{Ker } \phi| = 5$. So, $|\text{Ker } \phi| = 6$. The only subgroup of Z_{30} of order 6 is $\langle 5 \rangle$.

22. Let $|\phi(g)| = 8$. By Theorem 10.1 part 3, $|g| = 8k$. Then $|g^k| = 8$. To generalize replace 8 by n.

23. $|\phi^{-1}(H)| = |H||\text{Ker } \phi|$.

24.
a. Let $\phi(1) = k$. Then $\phi(7) = 7k \bmod 15 = 6$ so that $k = 3$ and $\phi(x) = 3x$.
b. $\langle 3 \rangle$.
c. $\langle 5 \rangle$.
d. $4 + \langle 5 \rangle$.

25. To define a homomorphism from Z_{20} onto Z_{10} we must map 1 to a generator of Z_{10}. Since there are four generators of Z_{10} we have four homomorphisms. (Once we specify that 1 maps to an element a, the homomorphism is $x \to xa$.) To define a homomorphism from Z_{20} to Z_{10} we can map 1 to any element of Z_{10}. (Be careful here, these mappings are well defined only because 10 divides 20.)

26. There are four: $x \to (x \bmod 2, 0)$; $x \to (0, x \bmod 2)$;
$x \to (x \bmod 2, x \bmod 2)$; $x \to (0, 0)$.

27. If ϕ is a homomorphism from Z_n to Z_n with $\phi(1) = k$, then by property 2 of Theorem 10.1, $\phi(x) = kx$. Moreover, for each k with $0 \leq k \leq n-1$, the mapping $\phi(x) = kx$ is a homomorphism.

28. Ker $\phi = A_4$. The trivial homomorphism and the one given in Example 11 are the only homomorphisms. To verify this observe that by Theorem 10.3, $|\text{Ker } \phi| = 12$ or 24. If $|\text{Ker } \phi| = 12$ one possibility for Ker ϕ is A_4. If H is another one not A_4, then since A_4 is normal in S_4, HA_4 is a subgroup of S_4 of order greater than 12. So, by Theorem 7.2, $|HA_4| = 12 \cdot 12/|H \cap A_4| = 24$, which implies that $H \cap K$ is a subgroup of A_4 of order 6. But Example 5 in Chapter 7 rules that out.

29. Say the kernel of the homomorphism is K. By Theorem 10.3, $G/K \approx Z_{10}$. So, $|G| = 10|K|$. In Z_{10}, let $\overline{H} = \langle 2 \rangle$. By properties 5, 7, and 9 of Theorem 10.2, $\phi^{-1}(\overline{H})$ is a normal subgroup of G of order $2|K|$. So, $\phi^{-1}(\overline{H})$ has index 2. To show that there is a subgroup of G of index 5, use the same argument with $\overline{H} = \langle 5 \rangle$. If there is a homomorphism from a finite group G onto Z_n, then the same argument shows that G has a normal subgroup of index d for any divisor D of n.

30. $Z_6 \oplus Z_2$ has normal subgroups of orders 1, 2, 3, 4,6, and 12. So by parts 5 and 9 of Theorem 10.2, G has normal subgroups of orders 5, 10, 15, 20, 30, and 60.

31. Suppose that the correspondence from Z_m to Z_n given by $x \to kx$ is a homomorphism. Then, in Z_n, we have $0 = \phi(0) = \phi(m) = mk$. So, n divides mk. Now suppose n divides mk. We first show that ϕ is well-defined. That is, if $a = b \bmod m$, we must show that $\phi(a) = \phi(b) \bmod n$. From $a = b \bmod m$, we have $a - b = mt$ for some integer t. Then, $\phi(a) - \phi(b) = ka - kb = k(a - b) = kmt = 0$ mod n. Finally, observe that for a and b in Z_m, we have that in Z_n, $\phi(a + b) = k(a + b) = ka + kb = \phi(a) + \phi(b)$.

32. $\{7, 12, 17, 22, 27, 2\}; \{14, 19, 24, 29, 4, 9\}; \{21, 26, 1, 6, 11, 16\}$.

33. By property 6 of Theorem 10.1, $\phi^{-1}(11) = 11\text{Ker } \phi = \{11, 19, 27, 3\}$.

34. First observe from Table 5.1 that A_4 has three elements of order 2 and eight of order 3. Thus, by property 3 of Theorem 10.1 the orders of the elements of G must be 1, 2, or 3. So, since G is cyclic, $|G| = 1, 2$, or 3. If $|G| = 2$, then $|\text{Ker } \phi| = 6$. This means that for every element α of order 3, we have that $|\phi(\alpha)|$ must divide both 2 and 3 and therefore $\phi(\alpha) = \epsilon$. But then Ker ϕ would contain all 8 elements of A_4 of order 3, which is a contradiction.

35. $\phi((a, b) + (c, d)) = \phi((a + c, b + d)) = (a + c) - (b + d) = (a - b) + (c - d) = \phi((a, b)) + \phi((c, d))$. Ker $\phi = \{(a, a) \mid a \in Z\}$. $\phi^{-1}(3) = \{(a + 3, a) \mid a \in Z\}$.

36. $4a - 4b$.

37. Consider the mapping ϕ from C^* onto R^+, given by $\phi(x) = |x|$. (Recall from Chapter 0 that $|a + bi| = \sqrt{a^2 + b^2}$.) By straightforward algebra we have $|xy| = |x||y|$. Thus ϕ is a homomorphism with Ker $\phi = H$. So, by Theorem 10.3, C^*/H is isomorphic to R^+.

38. Ker γ = Ker $\alpha \oplus$ Ker β.

39. $\phi(xy) = (xy)^6 = x^6y^6 = \phi(x)\phi(y)$. Ker $\phi = \langle \cos 60^\circ + i \sin 60^\circ \rangle$.

40. $\langle 12 \rangle$; $\langle 12 \rangle$; in general, the kernel is $\langle \text{lcm}(m, n) \rangle$.

41. Since $\phi(e) = e = e^{-2}$, $e \in H$. If $a \in H$, then $\phi(ab) = \phi(a)\phi(b) = a^{-2}b^{-2} = (ab)^{-2} \in H$. Also, $\phi(a^{-1}) = \phi(a)^{-1} = (a^{-2})^{-1} = (a^{-1})^{-2} \in H$. If $\phi(x) = x^3$ and $a \in H$, then $\phi(a) = a^3 = a^{-2}$ implies that $a^5 = e$. Thus, $|a| = 5$ or 1.

42. Because $|D_6/\text{Ker } \phi| = |D_3| = 6$, we know $|\text{Ker } \phi| = 2$. Since $\{R_0, R_{180}\}$ is the only normal subgroup of D_6 of order 2, it is Ker ϕ.

43. Property 2 of Theorem 10.2 handles the 2^m case. Suppose that there is a homomorphism from $G = Z_{2^m} \oplus Z_{2^n}$ onto $Z_2 \oplus Z_2 \oplus Z_2$ where m and n are at least 1 and let H be the kernel. We may assume that $m + n \geq 3$. Then $|H| = 2^{m+n-3}$. Because every nonidentity element of G/H has order 2, we know that $((1,0)H)^2 = (2,0)H = H$ and $((0,1)H)^2 = (0,2)H = H$. This means that $H_1 = \langle(2,0)\rangle$ and $H_2 = \langle(0,2)\rangle$ are subgroups of H. Then H_1H_2 is also a subgroup of H. But $|H_1H_2| = 2^{m-1} \cdot 2^{n-1} = 2^{m+n-2}$ exceeds $|H| = 2^{m+n-3}$. This argument works for any prime p.

44. If there were a homomorphism ϕ from S_3 onto Z_3, then ϕ would be a 2-1 mapping. But, because Z_3 has 2 elements of order 3, S_3 would have to have four elements of order 3, which it does not. For S_4 the mapping would be 8-1. Then S_4 would have to have 16 elements of order 3 to map only the 2 elements in Z_3 of order 3. But it does not.

45. Let H be the normal subgroup of order 4 defined in Example 9. Then S_4/H is a group of order 6 but has no element of order 6 (because S_4 does not have one). So, it follows from Theorem 7.3, S_4/H is isomorphic S_3.

46. Let ϕ be a homomorphism of $Z_m \oplus Z_n$. Then $\phi(Z_m \oplus Z_n) = \phi(\langle(1,0)\rangle \oplus \langle(0,1)\rangle) = \phi(\langle(1,0)\rangle) \oplus \phi(\langle(0,1)\rangle)$. Now use property 3 of Theorem 10.1.

47. It follows from Exercise 11 in Chapter 0 that the mapping ϕ from $U(st)$ to $U(s)$ given by $\phi(x) = x \bmod s$ is a homomorphism. Since Ker $\phi = U_s(st)$ we have by Theorem 10.3 that $U(st)/U_s(st)$ is isomorphic to a subgroup of $U(s)$. To see that ϕ is onto, note that it follows from Theorem 8.3 that $|U(st)/U_s(st)| = |U(st)|/|U_s(st)| = |U(s) \oplus U(t)|/|U(t)| = |U(s)||U(t)|/|U(t)| = |U(s)|$.

48. Observe that every element can be written in the form $s_1^{n_1} s_2^{n_2} \ldots s_m^{n_m}$ where $s_1 s_2, \ldots, s_m \in S$ and the exponents are integers. So, $\phi(g) = \phi(s_1)^{n_1} \phi(s_2)^{n_2} \cdots \phi(s_m)^{n_m} \in \phi(S)$.

49. Consider the mapping ϕ from K to KN/N given by $\phi(k) = kN$. Since $\phi(kk') = kk'N = kNk'N = \phi(k)\phi(k')$ and $kN \in KN/N$, ϕ is a homomorphism. Moreover, Ker $\phi = K \cap H$. So, by Theorem 10.3, $K/(K \cap N) \approx KN/N$.

50. Show that the mapping from G/N to G/M given by $gN \to gM$ is an onto homomorphism with kernel M/N.

51. Since the eight elements of A_4 of order 3 must map to an element of order that divides 3, by Lagrange's Theorem, each of them must map to the identity. But then the kernel has at least 8 elements and its order and divides 12. So, the kernel has order 12.

52. $U_k(n)$ is the kernel.

53. Let N be a normal subgroup of D_4. By Lagrange's Theorem, the only possibilities for $|N|$ are 1, 2, 4, and 8. By Theorem 10.4, the homomorphic images of D_4 are the same as the factor groups D_4/N of D_4. When $|N| = 1$, we know $N = \{e\}$ and $D_4/N \approx D_4$. When $|N| = 2$, then $N = \{R_0, R_{180}\}$, since this is the only normal subgroup of D_4 of order 2, and $D_4/N \approx Z_2 \oplus Z_2$ because D_4/N is a group of order 4

with three elements of order 2. When $|N| = 4$, $|D_4/N| = 2$ so $D_4/N \approx Z_2$. When $|N| = 8$, we have $D_4/N \approx \{e\}$.

54. Use Theorem 10.4 and part 3 of Theorem 10.1.

55. It is divisible by 10. In general, if Z_n is the homomorphic image of G, then $|G|$ is divisible by n.

56. It is divisible by 30. In general, the order of G is divisible by the least common multiple of the orders of all its homomorphic images.

57. It is infinite. Z is an example.

58. Let A be the coefficient matrix of the system. If A is an $n \times m$ matrix, then matrix multiplication by A is a homomorphism from $\mathbf{R}^m$ into $\mathbf{R}^n$ whose kernel is S.

59. Let γ be a natural homomorphism from G onto G/N. Let $\overline{H}$ be a subgroup of G/N and let $\gamma^{-1}(\overline{H}) = H$. Then H is a subgroup of G and $H/N = \gamma(H) = \gamma(\gamma^{-1}(\overline{H})) = \overline{H}$.

60. Use Theorem 10.1, part 2.

61. The mapping $g \to \phi_g$ is a homomorphism with kernel $Z(G)$.

62. **a.** Since $4 = |Z_2 \oplus Z_2|$ does not divide $|D_5|$, there are none.

    b. There are four. In addition to the trivial homomorphism, we can map all rotations to the identity and all reflections to any one of the three elements of order 2.

63. Since $(f+g)(3) = f(3) + g(3)$, the mapping is a homomorphism. The kernel is the set of elements in $Z[x]$ whose graphs pass through the point $(3, 0)$. 3 can be replaced by any integer.

64. For the first part, use trig identities. The kernel is $\langle 2\pi \rangle$.

65. Let g belong to G. Since $\phi(g)$ belongs to $Z_2 \oplus Z_2 = \langle(1,0)\rangle \cup \langle(0,1)\rangle \cup \langle(1,1)\rangle$, it follows that $G = \phi^{-1}(\langle(1,0)\rangle) \cup \phi^{-1}(\langle(0,1)\rangle) \cup \phi^{-1}(\langle(1,1)\rangle)$. Moreover, each of these three subgroups is proper since ϕ is onto and each is normal by property 8 of Theorem 10.2.

66. Try $g \to (gH, gK)$.

67. Map (a, b) to $(a \bmod 4, b)$.

68. Since $\phi(Z(D_{12})) \subseteq Z(D_3) = \{R_0\}$, we know $\phi(R_{180}) = R_0$.

69. It fails because 5 does divide $|\text{Aut}(Z_{11})| = 10$.

70. The argument in Example 18 shows that G has a unique subgroup H of order 11 and a subgroup K of order 5. If K was a normal subgroup, then both H and K would be normal (see Example 7 in Chapter 9) and we would have $G = H \times K \approx H \oplus K \approx Z_{11} \oplus Z_5 \approx Z_{55}$, which is a contradiction. Let $a \in H$ and $a \neq e$. Since $K \subseteq N(K)$ and $N(K) \neq G$, we know a is not in $N(K)$. We claim that the subgroups $a^i K a^{-i}$ for $i = 0, 1, 2, \ldots, 10$ are distinct. To verify this, suppose that $i, j \in \{0, 1, 2, \ldots, 10\}$ with $i > j$ and $a^i K a^{-i} = a^j K a^{-j}$. Then $a^{i-j} K a^{-i-j} = K$. This means that $a^{i-j} \in N(K)$. But then $a \in N(K)$ because $a^{i-j} \neq e$ and $|a| = 11$ implies that $\langle a^{i-j} \rangle = \langle a \rangle$.

71. Mimic Example 18.

72. There are no homomorphisms from Z onto S_3 since the image of a cyclic group must be cyclic. For each element x in S_3 the mapping from Z to S_3 given by $\phi(n) = x^n$ is a homomorphism. It follows from part 2 of Theorem 10.2 that there are no others.

73. Let ϕ be a homomorphism from S_3 to Z_n. Since $|\phi(S_3)|$ must divide 6, we have that $|\phi(S_3)| = 1, 2, 3,$ or 6. In the first case, ϕ maps every element to 0. If $|\phi(S_3)| = 2$, then n is even and ϕ maps the even permutations to 0, and the odd permutations to $n/2$. The case that $|\phi(S_3)| = 3$ cannot occur because it implies that Ker ϕ is a normal subgroup of order 2 whereas S_3 has no normal subgroup of order 2. The case that $|\phi(S_3)| = 6$ cannot occur because it implies that ϕ is an isomorphism from a non-Abelian group to an Abelian group.

74. Suppose that ϕ is a homomorphism and $m = nq + r$ where $1 \leq r < n$. Then $0 = \phi(0) = \phi(nq + r) = \phi(nq) + \phi(r) = r \neq 0$. This contradiction shows that for ϕ to be a homomorphism it is necessary that n divides m. If n divides m, then ϕ is well-defined and is operation preserving from Exercise 11 of Chapter 0.

75. $\phi(zw) = z^2w^2 = \phi(z)\phi(w)$. Ker $\phi = \{1, -1\}$ and, because ϕ is onto $\mathbf{C}^*$, we have by Theorem 10.3, that $\mathbf{C}^*/\{1, -1\}$ is isomorphic to $\mathbf{C}^*$. If $\mathbf{C}^*$ is replaced by $\mathbf{R}^*$, we have that ϕ is onto $\mathbf{R}^+$ and by Theorem 10.3, $\mathbf{R}^*/\{1, -1\}$ is isomorphic to $\mathbf{R}^+$.

76. p^2. To verify this, note that for any homomorphism ϕ from $Z_p \oplus Z_p$ into Z_p we have $\phi(a, b) = a\phi(1, 0) + b\phi(0, 1)$. Thus we need only count the number of choices for $\phi(1, 0)$ and $\phi(0, 1)$. Since p is prime, we may let $\phi(1, 0)$ be any element of Z_p. The same is true for $\phi(0, 1)$.

77. If a homomorphism ϕ maps 1 to k, then $\phi(x) = kx$. Thus, by property 3 of Theorem 10.1, $|a|$ must be a divisor of 5. This gives us that the possibilities for k are 0, 3, 6, 9, and 12. These five possibilities for k can be checked with one argument by noting that they all have the form $3s$ and proceed as in Example 11.

78. Say ϕ is such a homomorphism. Then, in all three cases, for any x we have $\phi(x) = \phi(x/2 + x/2) = \phi(x/2) + \phi(x/2) = 2\phi(x/2) = 0$ mod 2. This proof works for Z_n where n is even, since $\phi(x) = n(x/n) = 0 \mod n$.

79. First note that by by the Classification of Group of order $2p$, where p is an odd prime (Theorem 7.3), we have that S_3 is isomorphic to D_3. By properties of cosets, $|S_4/H| = 6$ and by Theorem 7.3, we have that S_4/H is isomorphic to Z_6 or D_3. Because the largest order of any element in S_4 is 4, S_4/H is isomorphic to D_3.

80. First note that the orders of the elements in S_4 are 1, 2, 3, and 4 and that the number of elements of order 3 is $4 \cdot 3 \cdot 2/3 = 8$. This means that the number of elements of orders 1, 2 or 4 is 16. Suppose that H is a normal subgroup of S_4 of order 6. Then there is a homomorphism $\phi$from S_4/H onto a group of order 4. But, for any element α in S_4 of order 3 we know that $|\phi(\alpha)|$ must divide both 3 and 4. So, H would contain all 8 elements of order 3.
Now suppose that S_4 has a normal subgroup H of order 8. Then there is a homomorphism ϕ from S_4/H onto a group of order 3. But, for any element α in S_4 of order 1, 2 or 4 we know that $|\phi(\alpha)|$ must divide both 3 and 4. So, $|\phi(\alpha)| = 1$ and H would contain all 16 elements of orders 1, 2, or 4.

81. We know that $|\phi(1)|$ must divide 24 and 9. So, $|\phi(1)| = 1$ or 3. This tells us that $\phi(1) = 0, 3,$ or 6.

CHAPTER 11

Fundamental Theorem of Finite Abelian Groups

1. $n = 4$
   Z_4, $Z_2 \oplus Z_2$

2. $n = 8$; $Z_8, Z_4 \oplus Z_2, Z_2 \oplus Z_2 \oplus Z_2$

3. $n = 36$
   $Z_9 \oplus Z_4$, $Z_3 \oplus Z_3 \oplus Z_4$, $Z_9 \oplus Z_2 \oplus Z_2$, $Z_3 \oplus Z_3 \oplus Z_2 \oplus Z_2$

4. Order 2: 1, 3, 3, 7; order 4: 2, 4, 12, 8

5. The only Abelian groups of order 45 are Z_{45} and $Z_3 \oplus Z_3 \oplus Z_5$. In the first group, $|3| = 15$; in the second one, $|(1, 1, 1)| = 15$. $Z_3 \oplus Z_3 \oplus Z_5$ does not have an element of order 9.

6. $Z_{27} \oplus Z_4$; $Z_{27} \oplus Z_2 \oplus Z_2$.

7. In order to have exactly four subgroups of order 3, the group must have exactly 8 elements of order 3. When counting elements of order 3 we may ignore the components of the direct product that represent the subgroup of order 4 since their contribution is only the identity. Thus, we examine Abelian groups of order 27 to see which have exactly 8 elements of order 3. By Theorem 4.4, Z_{27} has exactly 2 elements of order 3; $Z_9 \oplus Z_3$ has exactly 8 elements of order 3 since for $|(a, b)| = 3$ we can choose $|a| = 1$ or 3 and $|b| = 1$ or 3, but not both $|a|$ and $|b|$ of order 1; in $Z_3 \oplus Z_3 \oplus Z_3$, every element except the identity has order 3. So, the Abelian groups of order 108 that have exactly four subgroups of order 3 are $Z_9 \oplus Z_3 \oplus Z_4$ and $Z_9 \oplus Z_3 \oplus Z_2 \oplus Z_2$. The subgroups of $Z_9 \oplus Z_3 \oplus Z_4$ of order 3 are $\langle(3, 0, 0)\rangle, \langle(0, 1, 0)\rangle, \langle(3, 1, 0)\rangle$ and $\langle(3, 2, 0)\rangle$. The subgroups of $Z_9 \oplus Z_3 \oplus Z_2 \oplus Z_2$ of order 3 are $\langle(3, 0, 0, 0)\rangle, \langle(0, 1, 0, 0)\rangle, \langle(3, 1, 0, 0)\rangle$ and $\langle(3, 2, 0, 0)\rangle$.

8. $Z_3 \oplus Z_3 \oplus Z_3 \oplus Z_4$; $Z_3 \oplus Z_3 \oplus Z_3 \oplus Z_2 \oplus Z_2$.

9. Elements of order 2 are determined by the factors in the direct product that have order a power of 2. So, we need only look at Z_8, $Z_4 \oplus Z_2$ and $Z_2 \oplus Z_2 \oplus Z_2$. By Theorem 4.4, Z_8 has exactly one element of order 2; $Z_4 \oplus Z_2$ has exactly three elements of order 2; $Z_2 \oplus Z_2 \oplus Z_2$ has exactly 7 elements of order 2. So, $G \approx Z_4 \oplus Z_2 \oplus Z_3 \oplus Z_5$.

10. $Z_8 \oplus Z_9 \oplus Z_5$; $Z_4 \oplus Z_2 \oplus Z_9 \oplus Z_5$; $Z_2 \oplus Z_2 \oplus Z_2 \oplus Z_9 \oplus Z_5$; $Z_8 \oplus Z_3 \oplus Z_3 \oplus Z_5$; $Z_4 \oplus Z_2 \oplus Z_3 \oplus Z_3 \oplus Z_5$; $Z_2 \oplus Z_2 \oplus Z_2 \oplus Z_3 \oplus Z_3 \oplus Z_5$.

11. By the Fundamental Theorem, any finite Abelian group G is isomorphic to some direct product of cyclic groups of prime-power order. Now go across the direct product and, for each distinct prime you have, pick off the largest factor of that prime-power. Next, combine all of these into one factor (you can do this, since their orders are relatively prime). Let us call the order of this new factor n_1. Now repeat this process with the remaining original factors and call the order of the resulting

factor n_2. Then n_2 divides n_1, since each prime-power divisor of n_2 is also a prime-power divisor of n_1. Continue in this fashion. Example: If

$$G \approx Z_{27} \oplus Z_3 \oplus Z_{125} \oplus Z_{25} \oplus Z_4 \oplus Z_2 \oplus Z_2,$$

then

$$G \approx Z_{27 \cdot 125 \cdot 4} \oplus Z_{3 \cdot 25 \cdot 2} \oplus Z_2.$$

Now note that 2 divides $3 \cdot 25 \cdot 2$ and $3 \cdot 25 \cdot 2$ divides $27 \cdot 125 \cdot 4$.

12. By the corollary to the Fundamental Theorem of Finite Abelian Groups, the given group has a subgroup of order 10. But this group must be isomorphic to $Z_2 \oplus Z_5 \approx Z_{10}$.

13. $Z_2 \oplus Z_2$

14. If G is an Abelian group of order n and m is a divisor of n, then G has a cyclic subgroup of order m if m is square-free (i.e., each prime factor of m occurs to the 1st power only).

15. **a.** 1 **b.** 1 **c.** 1 **d.** 1 **e.** 1 **f.** There is a unique Abelian group of order n if and only if n is not divisible by the square of any prime.

16. **a.** same **b.** same **c.** same **d.** same
e. twice as many of order m compared with the number of order n

17. This is equivalent to asking how many Abelian groups of order 16 have no element of order 8. From the Fundamental Theorem of Finite Abelian Groups, the only choices are $Z_4 \oplus Z_4$, $Z_4 \oplus Z_2 \oplus Z_2$, and $Z_2 \oplus Z_2 \oplus Z_2 \oplus Z_2$.

18. 5^n

19. The symmetry group is $\{R_0, R_{180}, H, V\}$. Since this group is Abelian and has no element of order 4, it is isomorphic to $Z_2 \oplus Z_2$.

20. Consider every possible isomorphism class one by one and show each has the desired subgroup. For instance, in $Z_2 \oplus Z_2 \oplus Z_2 \oplus Z_{27} \oplus Z_5$ the subgroup is

$$\{(a, b, 0, c, d) \mid a, b \in Z_2,\ c \in \langle 3 \rangle,\ d \in Z_5\}$$

21. Because the group is Abelian and has order 9, the only possibilities are Z_9 and $Z_3 \oplus Z_3$. Since Z_9 has exactly 2 elements of order 3 and 9, 16, and 22 have order 3, the group must be isomorphic to $Z_3 \oplus Z_3$.

22. Because of the Fundamental Theorem and Corollary 1 of Theorem 8.2, we may assume $|G|$ is a prime-power. Let x be an element of G of maximum order. Then for any y in G we have $|y|$ divides $|\langle x \rangle|$. Since $\langle x \rangle$ has a unique subgroup for each divisor of $|\langle x \rangle|$, it follows that $\langle y \rangle \subseteq \langle x \rangle$.

23. By the Corollary of Theorem 8.2, n must be square-free (no prime factor of n occurs more than once).

24. $n = p_1^2 p_2^2$ or $p_1^2 p_2^2 p_3 p_4 \cdots p_k$ where $k \geq 3$ and $p_1, p_2, \ldots, p_k$ are distinct primes.

25. Among the first 11 elements in the table, there are 9 elements of order 4. None of the other isomorphism classes has this many.

26. $Z_4 \oplus Z_2$; one internal direct product is $\langle 7\rangle \times \langle 17\rangle$. Some others are $\langle 7\rangle \times \langle 65\rangle$ and $\langle 23\rangle \times \langle 65\rangle$.

27. First observe that G is Abelian and has order 16. Now we check the orders of the elements. Since the group has 8 elements of order 4 and 7 of order 2, it is isomorphic to $Z_4 \oplus Z_2 \oplus Z_2$. One internal direct product is $\langle 7\rangle \times \langle 101\rangle \times \langle 199\rangle$.

28. $Z_2 \oplus Z_2 \oplus Z_3$; One internal direct product is $\langle 19\rangle \times \langle 26\rangle \times \langle 31\rangle$.

29. Since $|\langle(2,2)\rangle| = 8$, we know $|(Z_{16} \oplus Z_{16})/\langle(2,2)\rangle| = 32$. Because the largest order of any element in $Z_{16} \oplus Z_{16}$ is 16, the same is true about the factor group. Then observing that $|(1,0) + \langle(2,2)\rangle| = 16$, we know that the isomorphism class is $Z_{16} \oplus Z_2$.

30. $Z_4 \oplus Z_2 \oplus Z_4$

31. Since Z_9 has exactly 2 elements of order 3, once we choose 3 nonidentity elements we will either have at least one element of order 9 or 3 elements of order 3. In either case we have determined the group. The Abelian groups of order 18 are $Z_9 \oplus Z_2 \approx Z_{18}$ and $Z_3 \oplus Z_3 \oplus Z_2$. By Theorem 4.4, Z_{18} group has 6 elements of order 18, 6 elements of order 9, 2 of order 6, 2 of order 3, 1 of order 2, and 1 of order 1. $Z_3 \oplus Z_3 \oplus Z_2$ has 8 elements of order 3, 8 of order 6, 1 of order 2, and 1 of order 1. The worst-case scenario is that at the end of 5 choices we have selected 2 of order 6, 2 of order 3, and 1 of order 2. In this case we still have not determined which group we have. But the sixth element we select will give us either an element of order 18 or 9, in which case we know the group Z_{18} or a third element of order 6 or 3, in which case we know the group is $Z_3 \oplus Z_3 \oplus Z_2$.

32. The element of order 8 rules out all but Z_{16} and $Z_8 \oplus Z_2$ and two elements of order 2 precludes Z_{16}.

33. If $a^2 \neq b^2$, then $a \neq b$ and $a \neq b^3$. It follows that $\langle a\rangle \cap \langle b\rangle = \{e\}$. Then $G = \langle a\rangle \times \langle b\rangle \approx Z_4 \oplus Z_4$.

34. Observe that the subgroup $H = \{x \in G \mid x^2 = e\}$ consist exactly of the elements of G of order 2 and e. So, it follows for the corollary of Theorem 11.1 that $|H|$ has the form 2^n for some positive integer n and that the number of elements of order 2 is $2^n - 1$. This argument shows that for any finite Abelian group the number of elements of order p, where is a prime, has the form $p^n - 1$.

35. By Theorem 11.1, we can write the group in the form $Z_{{p_1}^{n_1}} \oplus Z_{{p_2}^{n_2}} \oplus \cdots \oplus Z_{{p_k}^{n_k}}$ where each p_i is an odd prime. By Theorem 8.1 the order of any element $(a_1, a_2, \ldots, a_k) = \text{lcm}(|a_1|, |a_2|, \ldots, |a_k|)$. And from Theorem 4.3 we know that $|a_i|$ divides $p_i^{n_i}$, which is odd.

36. $Z_2 \oplus Z_2 \oplus \cdots \oplus Z_2$ (n terms).

37. By Theorem 7.2 we have, $|\langle a\rangle K| = |a||K|/|\langle a\rangle \cap K| = |a||K| = |\overline{a}||\overline{K}|p = |\overline{G}|p = |G|$.

38. If $|G| = p^n$, use the Fundamental Theorem and Theorem 9.6. If every element has order a power of p, use the corollary to the Fundamental Theorem.

39. By the Fundamental Theorem of Finite Abelian Groups, it suffices to show that every group of the form $Z_{p_1^{n_1}} \oplus Z_{p_2^{n_2}} \oplus \cdots \oplus Z_{p_k^{n_k}}$ is a subgroup of a U-group. Consider first a group of the form $Z_{p_1^{n_1}} \oplus Z_{p_2^{n_2}}$ (p_1 and p_2 need not be distinct). By Dirichlet's Theorem, for some s and t there are distinct primes q and r such that $q = tp_1^{n_1} + 1$ and $r = sp_2^{n_2} + 1$. Then $U(qr) = U(q) \oplus U(r) \approx Z_{tp_1^{n_1}} \oplus Z_{sp_2^{n_2}}$, and this

latter group contains a subgroup isomorphic to $Z_{p_1^{n_1}} \oplus Z_{p_2^{n_2}}$. The general case follows in the same way.

40. Observe that
$\text{Aut}(Z_2 \oplus Z_3 \oplus Z_5) \approx \text{Aut}(Z_{30}) \approx U(30) \approx U(2) \oplus U(3) \oplus U(5) \approx Z_2 \oplus Z_4$.

41. It follows from Exercise 4 of Chapter 8 and Theorem 9.6 that if D_4 could be written in the form $\langle a \rangle \times K$ where $|a| = 4$, it would be Abelian.

42. If is an Abelian group G of order 2^n is not cyclic, then by Theorem 11.1, G is isomorphic to a group of the form $Z_{2^{m_1}} \oplus Z_{2^{m_2}} \oplus \cdots \oplus Z_{2^{m_k}}$ where $k \geq 2$ and each m_i is at least 1. But the subgroup $Z_{2^{m_1}} \oplus Z_{2^{m_2}} \oplus \{0\} \cdots \oplus \{0\}$ has three elements of order 2. The other half follows from Theorem 4.4. In general, an Abelian group of order p^n is cyclic if and only if it has exactly $p-1$ elements of order p.

43. If G has an element of order greater than 2, then $\phi(x) = x^{-1}$ is a non-trivial automorphism of G (see Exercise 12 of Chapter 6). If not, then $|G| = 2^n$ and is G isomorphic to $Z_2 \oplus Z_2 \oplus \cdots \oplus Z_2$ (n terms). Then
$\phi(x_1, x_2, x_3, \ldots, x_n) = \phi(x_2, x_1, x_3, \ldots, x_n)$ is an automorphism of G.

44. This follows directly from Exercise 11. An example for the second part is D_3.

45. By Theorem 11.1 and Corollary 1 of Theorem 8.2, it suffices to do the case where $|G| = p^m$ and p is prime. By Theorem 11.1, if G is not cyclic, then G is isomorphic to a group of the form $Z_{p^{m_1}} \oplus Z_{p^{m_2}} \oplus \cdots \oplus Z_{p^{m_k}}$ where $k \geq 2$ and each m_i is at least 1. But then, by Theorem 4.3, G has a subgroup of the form $Z_p \oplus Z_p \oplus \cdots \oplus Z_p$ of order p^k and every element of this subgroup is a solution to $x^p = e$.

46. By the Fundamental Theorem of Finite Abelian groups, G can be written as an internal direct product of the form $A \times B$ where $|A|$ is a power of 2. Then, for any nonidentity elements $a \in A$ and any element $b \in B$ we know that ab has even order. This gives us $|H| \geq |(|A|-1)B| \geq (|A|/2)|B| = |A||B|/2 = |G|/2$ elements of even order. Since H also contains the identity, $H = G$.

47. First, observe by direct calculations we have that $|8| = |12| = |18| = |21| = |27| = 4$. Since for all x in G we also have $|x| = |-x| = |65 - x|$, we know that G has at least 10 elements of order 4. By Theorem 4.4, Z_{18} has only 2 elements of order 4 and by Theorem 8.1 $Z_2 \oplus Z_2 \oplus Z_2 \oplus Z_2$ has none, so these two groups are eliminated. Finally, arguing as in Examples 5 and 6 in Chapter 8 we know that $Z_8 \oplus Z_2$ has only 4 elements of order 4 and $Z_4 \oplus Z_2 \oplus Z_2$ has only 8. So, $G \approx Z_4 \oplus Z_4$.

48. $Z_{{p_1}^{n_1}} \oplus Z_{{p_2}^{n_2}} \oplus \cdots \oplus Z_{{p_k}^{n_k}}$ where the p_i terms are distinct primes.

CHAPTER 12

Introduction to Rings

1. For any $n > 1$, the ring $M_2(Z_n)$ of 2×2 matrices with entries from Z_n is a finite noncommutative ring. The set $M_2(2Z)$ of 2×2 matrices with even integer entries is an infinite noncommutative ring that does not have a unity.

2. 6 is the unity; $2 \cdot 8 = 6$, $4 \cdot 4 = 6$.

3. In **R**, $\{n\sqrt{2} | \ n \in Z\}$ is a subgroup but not a subring. Another example is the ring $M_2(Z)$ and the subgroup of all elements with the entry 0 in the upper left corner.

4. In Z_4, $2x = 2$ has solutions 1 and 3. In a group, $x = a^{-1}b$.

5. **a** and **b**

6. Consider Z_6 or Z_{10}.

7. First observe that every nonzero element a in Z_p has a multiplicative inverse a^{-1}. For part a, if $a \neq 0$, then $a^2 = a$ implies that $a^{-1}a^2 = a^{-1}a$ and therefore $a = 1$. For part b, if $a \neq 0$, then $ab = 0$ implies that $b = a^{-1}(ab) = a^{-1}0 = 0$. For part c, $ab = ac$ implies that $a^{-1}(ab) = a^{-1}(ac)$. So $b = c$.

8. Consider $aba = aba$.

9. If a and b belong to the intersection, then they belong to each member of the intersection. Thus $a - b$ and ab belong to each member of the intersection. So, $a - b$ and ab belong to the intersection.

10. Observe that all the sets in the examples are closed under subtraction and multiplication.

11. Rule 3: $0 = 0(-b) = (a + (-a))(-b) = a(-b) + (-a)(-b) = -(ab) + (-a)(-b)$. So, $ab = (-a)(-b)$.
Rule 4: $a(b - c) = a(b + (-c)) = ab + a(-c) = ab + (-(ac)) = ab - ac$.
Rule 5: By Rule 2, $(-1)a = 1(-a) = -a$.
Rule 6: By Rule 3, $(-1)(-1) = 1 \cdot 1 = 1$.

12. If $c = db$, then $c = d(a^{-1})ab = (da^{-1})ab$.
If $c = (ab)d$, then $c = (ad)b$.

13. Let S be any subring of Z. By definition of a ring, S is a subgroup under addition. By Theorem 4.3, $S = \langle k \rangle$ for some integer k.

14. Use induction.

15. If m or n is 0, the statement follows from part 1 of Theorem 12.1. For simplicity, for any integer k and any ring element x we will use kx instead of $k \cdot x$. Then for positive m and n, observe that
$(ma)(nb) = (a + a + \cdots + a) + (b + b + \cdots + b) = (ab + ab + \cdots + ab)$, where the terms $a + \cdots + a$, $b + b + \cdots + b$, and the last term have mn summands.

For the case that m is positive and n is negative, we first observe that nb means $(-b)+(-b)+\cdots+(-b)=(-n)(-b)$. So, $nb+(-n)b=((-b)+(-b)+\cdots+(-b))+(b+b+\cdots+b)=0$. Thus, $0=(ma)(nb+(-n)b)=(ma)(nb)+(ma)(-n)b=(ma)(nb)+m(-n)ab=(ma)(nb)+(-(mn))ab$. So, adding $(mn)ab$ to both ends of this string of equalities gives $(mn)ab=(ma)(nb)$.
For the case when m is negative and n is positive, just reverse the roles of m and n is the preceding argument. If both m and n are negative, note that $(ma)(nb)=((-a)+(-a)+\cdots+(-a))((-b)+(-b)+\cdots+(-b))=((-m)(-a))((-n)(-b))=(-m)(-n)((-a)(-b))=(mn)(ab)$.

16. Observe that $n\cdot(-a)+n\cdot a=0$.

17. From Exercise 15, we have $(n\cdot a)(m\cdot a)=(nm)\cdot a^2=(mn)\cdot a^2=(m\cdot a)(n\cdot a)$.

18. $r0=0$, so S is not empty. Let $a,b\in S$. Then $r(a-b)=ra-rb=0-0=0$. Also $r(ab)=(ra)b=0b=0$.

19. Let a,b belong to the center. Then $(a-b)x=ax-bx=xa-xb=x(a-b)$. Also, $(ab)x=a(bx)=a(xb)=(ax)b=(xa)b=x(ab)$.

20. $\langle 1\rangle\supset\langle 2\rangle\supset\langle 4\rangle\supset\langle 8\rangle\supset\cdots$.

21. $(x_1,\ldots,x_n)(a_1,\ldots,a_n)=(x_1,\ldots,x_n)$ for all x_i in R_i if and only if $x_ia_i=x_i$ for all x_i in R_i and $i=1,\ldots,n$, and $x_ia_i=x_i$ for all x_i in R_i if and only if x_i is a unity of R_i.

22. By the One-Step Test, we must show ab^{-1} is a unit wherever a and b are. But $ab^{-1}ba^{-1}=1$.

23. By observation ±1 and $\pm i$ are units. To see that there are no others, note that $(a+bi)^{-1}=\frac{1}{a+bi}=\frac{1}{a+bi}\frac{a-bi}{a-bi}=\frac{a-bi}{a^2+b^2}$. But $\frac{a}{a^2+b^2}$ is an integer only when $a^2+b^2=1$ and this holds only when $a=\pm1$ and $b=0$ or $a=0$ and $b=\pm1$.

24. Say e_i is the unity of R_i. Then

$$(a_1,\ldots,a_n)(b_1,\ldots,b_n)=(e_1,\ldots,e_n)$$

if and only if $a_ib_i=e_i$ for $i=1,\ldots,n$. That is, if and only if $a_i\in U(R_i)$.

25. Note that the only $f(x)\in Z[x]$ for which $1/f(x)$ is a polynomial with integer coefficients are $f(x)=1$ and $f(x)=-1$.

26. $\{f(x)=c\mid c\in\mathbf{R},\ c\neq0\}$.

27. If a is a unit, then $b=a(a^{-1}b)$.

28. In Z_6, $4\cdot2=2$; in Z_8, $3\cdot5=7$, in Z_{15}, $9\cdot3=12$.

29. Note that $(a+b)(a^{-1}-a^{-2}b)=1-a^{-1}b+ba^{-1}-a^{-2}b^2=1$.

30. $(1-a)(1+a+a^2+\cdots+a^{n-1})=1+a+a^2+\cdots+a^{n-1}-a-a^2-\cdots-a^n=1-a^n=1-0=1$.

31. $0^1=0$ so the set is nonempty. Let $a^m=0$ and $b^n=0$. We may assume that $m\geq n$. Then in the expansion of $(a-b)^{2m}$ each term has an expression of the form $a^{2m-i}b^i$. So when $i=0,1,\ldots,m$ we have $a^{2m-i}=0$ and when $i=m+1,m+2,m+m$ we have $b^i=0$. So, all terms in the expansion are 0. (This argument also works when the exponent of $(a-b)$ is $m+n-1$.) Finally, $(ab)^m=a^mr^m=0$.

32. We may assume that m is the smallest positive integer such that $a^m = 0$. If $m < n$, then $a = a^n = a^m a^{n-m} = 0a^{n-m} = 0$. If $m = n$, then $a = a^n = a^m = 0$. If $m > n$, observe that $0 = a^m = a^{m-n}a^n = a^{m-n}a = a^{m-n+1}$ and $m - n + 1 < m$, which is a contradiction to our assumption.
First alternate solution. Beginning with a and successively multiplying each prior term by a we get $a, a^2, a^3, \ldots a^n = a, a^2, \ldots, a^n = a, \ldots$, returning to a after every $n-1$ iterations. Thus, there exist arbitrarily large k such that $a^k = a$. So, for such $k \geq m$ we have $a = a^k = a^m a^{m-k} = 0$.
Second alternate solution. If $m < n$, then $a = a^n = a^m a^{n-m} = 0a^{n-m} = 0$. If $m = n$, then $a = a^n = a^m = 0$.

If $m > n$, observe that $a = a^n$ means that we can replace each of the n terms in a^n by a^n to obtain $a = (a^n)^n = a^{n^2}$. Then we can replace each of the n^2 terms in a^{n^2} $a == (a^{n^2})^{n^2} = a^{n^4}$. Continuing this process we eventually reach Case 1.

If $m > n$, observe that
$a = a^n = aa^{n-1} = a^n a^{n-1} = a^{2n-1} = aa^{2n-2} = a^n a^{2n-1} = 2^{3n-3}$. Continuing this for k iterations we have $a = a^{kn-(k-1)}$. Thus we eventually reach Case 1.

33. In $M_2(Z)$, let $a = \begin{bmatrix} 0 & 1 \\ 0 & 0 \end{bmatrix}$ and $b = \begin{bmatrix} 1 & 0 \\ 0 & 0 \end{bmatrix}$.

34. $ba = (ba)^n = baba \cdots = b(ab)a \cdots = 0$.

35. By inspection, R is closed under addition and multiplication. The elements $\begin{bmatrix} 0 & 1 \\ 0 & 0 \end{bmatrix}$ and $\begin{bmatrix} 0 & 1 \\ 0 & 1 \end{bmatrix}$ do not commute,

For the general case, use $m \times m$ matrices with the first $m-1$ columns all 0 and that last columns elements from Z_n.

36. For part **a** observe that $2x = (2x)^3 = 8x^3 = 8x$. So, $0 = 8x - 2x = 6x$. By part **a** we must have $|1|$ divides 6 so we need only check $n = 1, 2, 3$ and 6 and these have the desired property.

37. Observe that $-x = (-x)^4 = x^4 = x$.

38. Assume that $a^{2n} = a^2$. Then $a^{2(n+1)} = a^{2n}a^2 = a^2a^2 = a^4 = a^2$.

39. For Z_6 use $n = 3$. For Z_{10} use $n = 5$. Say $m = p^2t$ where p is a prime. Then $(pt)^n = 0$ in Z_m since m divides $(pt)^n$.

40. Say $k = ms$ and $k = nt$. Then $ka = (ms)a = m(sa)$ and $ka = (nt)a = n(ta)$ and therefore $ka \in mZ \cap nZ$. Now suppose $b \in mZ \cap nZ$. Then b is a common multiple of m and n. So, by Exercise 10 of Chapter 0, $b \in kZ$.

41. Every subgroup of Z_n is closed under multiplication.

42. No. The operations are different.

43. Since $ara - asa = a(r-s)a$ and $(ara)(asa) = ara^2sa = arsa$, S is a subring. Also, $a1a = a^2 = 1$, so $1 \in S$.

44. The set is not closed under multiplication.

45. Let $\begin{bmatrix} a & a-b \\ a-b & b \end{bmatrix}$ and $\begin{bmatrix} a' & a'-b' \\ a'-b' & b' \end{bmatrix} \in R$. Then $\begin{bmatrix} a & a-b \\ a-b & b \end{bmatrix} - \begin{bmatrix} a' & a'-b' \\ a'-b' & b' \end{bmatrix} = \begin{bmatrix} a-a' & (a-a')-(b-b') \\ (a-a')-(b-b') & b-b' \end{bmatrix} \in R$. Also, $\begin{bmatrix} a & a-b \\ a-b & b \end{bmatrix} \begin{bmatrix} a' & a'-b' \\ a'-b' & b' \end{bmatrix} = \begin{bmatrix} aa'+aa'-ab'-ba'+bb' & aa'-bb' \\ aa'-bb' & aa'-ab'-ba'+bb'+bb' \end{bmatrix}$ belongs to R.

46. The subring test is satisfied.

47. They satisfy the subring test but the multiplication is trivial. That is, the product of any two elements is zero.

48. The subring test is satisfied.

49. S is not a subring because $(1,0,1)$ and $(0,1,1)$ belong to S but $(1,0,1)(0,1,1) = (0,0,1)$ does not belong to S.

50. Say $n = 2m$. Then $-a = (-a)^n = (-a)^{2m} = ((-a)^2)^m = ((a)^2)^m = a^{2m} = a^n = a$.

51. Observe that $n \cdot 1 - m \cdot 1 = (n-m) \cdot 1$. Also, $(n \cdot 1)(m \cdot 1) = (nm) \cdot ((1)(1)) = (nm) \cdot 1$.

52. $2Z \cup 3Z$ contains 2 and 3, but not $2+3$.

53. $\{a_n(2/3)^n + a_{n-1}(2/3)^{n-1} + \cdots + a_1(2/3) \mid a_1, a_2, \ldots, a_n \in Z, n$ a positive integer$\}$. This set is a ring that contains $2/3$ and is contained in every ring that contains $2/3$. Alternate solution. $\{2n/3^m \mid n \in Z, m$ is a positive integer$\}$. This set is a ring that contains $2/3$ and is contained in every ring that contains $2/3$.

54. Observe that $1 = (a^{-1}b)c$ and $1 = b(ca^{-1})$.

55. $(a+b)(a-b) = a^2 + ba - ab - b^2 = a^2 - b^2$ if and only if $ba - ab = 0$.

56. First note that

$$a + b = (a+b)^2 = a^2 + ab + ba + b^2 = a + ab + ba + b$$

so that $0 = ab + ba$ or $-ab = ba$. Then observe that

$$-ab = (-ab)^2 = (ab)^2 = ab.$$

57. $Z_2 \oplus Z_2$; $Z_2 \oplus Z_2 \oplus \cdots$ (infinitely many copies).

58. $2x = 1$ has no solution in Z_4; $2x = 0$ has two solutions in Z_4; $x = a^{-1}(c-b)$ is the unique solution when a^{-1} exists.

59. If (a,b) is a zero-divisor in $R \oplus S$, then there is a $(c,d) \neq (0,0)$ such that $(a,b)(c,d) = (0,0)$. Thus $ac = 0$ and $bd = 0$. So, a or b is a zero-divisor or exactly one of a or b is 0. Conversely, if a is a zero-divisor in R, then there is a $c \neq 0$ in R such that $ac = 0$. In this case, $(a,b)(c,0) = (0,0)$. A similar argument applies if b is a zero-divisor. If $a = 0$ and $b \neq 0$, then $(a,b)(x,0) = (0,0)$ where x is any nonzero element in A. A similar argument applies if $a \neq 0$ and $b = 0$.

60. The inverse is $2x + 3$.

61. Fix some a in R, $a \neq 0$. Then there is a b in R such that $ab = a$. Now if $x \in R$ and $x \neq 0$, then there is an element c in R such that $ac = x$. Then $xb = acb = c(ab) = ca = x$. Thus b is the unity. To show that every nonzero element r of R has an inverse, note that since $rR = R$, there is an element s in R such that $rs = b$.

62. In Z_8, $2^2 = 4 = 6^2$ and $2^3 = 0 = 6^3$.

63. One solution is $R_0 = \langle 2^0 \rangle = Z$, $R_1 = \langle 2^1 \rangle$, $R_2 = \langle 2^2 \rangle, \ldots$.

64 Note that is r any element in a ring then $r = a(a^{-1}r)$.

65. In words, S is the set of all polynomials in π with the constant term 0. In symbols, $S = \{c_n\pi^n + c_{n-1}\pi^{n-1} + \cdots + c_1\pi \mid c_1, c_2, \ldots, c_n \in Z\}$. To see why S is the intersection of all subrings of $\mathbf{R}$ that contain π, let T be the intersection of all subrings of $\mathbf{R}$ that contain π. Then S is one subring in the intersection, so $T \subseteq S$, and, by definition, $S \subseteq T$.

66. Both parts can be done by solving $1 = (a^{-1} + b^{-1})x = 1$ for x. Doing so, we obtain $ab = ab(a^{-1} + b^{-1})x = (b + a)x = (a + b)x$, which gives $x = ab(a + b)^{-1}$.

67. Experimenting with $p = 3$ and $p = 5$ reveals that in both case $p + 1$ is the unity. So, we want to show that for all $x \in R$ we have that $(p + 1)x = x \bmod 2p$. Observe that because every x in R has 2 as a factor, $(p + 1)x = px + x = x \bmod 2p$.

68. R is closed under subtraction and multiplication.

69. R is not closed under multiplication.

70. Note that every element of R has the form $2A$ where A is in $M_2(Z)$. So, if $2A$ and $2B$ are in R, then $2A - 2B = 2(A - B)$ and $(2A)(2B) = 2(2AB)$ are in R. The argument works for any positive integer k.

CHAPTER 13

Integral Domains

1. For Example 1, observe that Z is a commutative ring with unity 1 and has no zero-divisors. For Example 2, note that $Z[i]$ is a commutative ring with unity 1 and no zero-divisors since it is a subset of **C**, which has no zero-divisors. For Example 3, note that $Z[x]$ is a commutative ring with unity $h(x) = 1$ and if $f(x) = a_n x^n + \cdots + a_0$ and $g(x) = b_m x^m + \cdots + b_0$ with $a_n \neq 0$ and $b_m \neq 0$, then $f(x)g(x) = a_n b_m x^{n+m} + \cdots + a_0 b_0$ and $a_n b_m \neq 0$. For Example 4, elements of $Z[\sqrt{2}]$ commute since they are real numbers; 1 is the unity; $(a + b\sqrt{2}) - (c + d\sqrt{2}) = (a - c) + (b - d)\sqrt{2}$ and $(a + b\sqrt{2})(c + d\sqrt{2}) = (ac + 2bd) + (bc + ad)\sqrt{2}$ so $Z[\sqrt{2}]$ is a ring; $Z[\sqrt{2}]$ has no zero-divisors because it is a subring of **R**, which has no zero-divisors. For Example 5, note that Z_p is closed under addition and multiplication and multiplication is commutative; 1 is the unity; in Z_p, $ab = 0$ implies that p divides ab. So, by Euclid's Lemma (see Chapter 0), we know that p divides a or p divides b. Thus, in Z_p, $a = 0$ or $b = 0$. For Example 6, if n is not prime, then $n = ab$ where $1 < a < n$ and $1 < b < n$. But then $a \neq 0$ and $b \neq 0$ while $ab = 0$. For Example 7, note that

$$\begin{bmatrix} 1 & 0 \\ 0 & 0 \end{bmatrix} \begin{bmatrix} 0 & 0 \\ 0 & 1 \end{bmatrix} = \begin{bmatrix} 0 & 0 \\ 0 & 0 \end{bmatrix}.$$

For Example 8, note that $(1, 0)(0, 1) = (0, 0)$.

2. Example 5

3. Let $ab = 0$ and $a \neq 0$. Then $ab = a \cdot 0$, so $b = 0$.

4. $2, 4, 5, 6, 8, 10, 12, 14, 15, 16, 18$. The zero-divisors and the units constitute a partition of Z_{20}.

5. Let $k \in Z_n$. If $\gcd(k, n) = 1$, then k is a unit. If $\gcd(k, n) = d > 1$, write $k = sd$. Then $k(n/d) = sd(n/d) = sn = 0$.

6. 2 in Z or x in $Z[x]$.

7. Let $s \in R$, $s \neq 0$. Consider the set $S = \{sr \mid r \in R\}$. If $S = R$, then $sr = 1$ (the unity) for some r. If $S \neq R$, then there are distinct r_1 and r_2 such that $sr_1 = sr_2$. In this case, $s(r_1 - r_2) = 0$.
Alternate solution. Let $a \in R$, $a \neq 0$, and $a \neq 1$. If there is some positive integer m such $a^m = 0$, let n be the least such integer. Then $aa^{n-1} = 0$ and a is a zero-divisor. If there is no such m consider, the infinite list $a, a^2, a^3, \ldots$. Since R is finite, we must have some distinct positive integers m and n $(m > n)$ with $a^m = a^n$. Then $a^m - a^n = a^n(a^{m-n} - 1) = 0$. If $a^{m-n} - 1 = 0$, a is a unit. If $a^{m-n} - 1 \neq 0$, a is a zero-divisor. To see what happens when the "finite" condition is dropped, note that in the ring of integers, 2 is neither a zero-divisor nor a unit.

8. Suppose that a is a zero-divisor and let $ab = 0$ for some $b \neq 0$. Then $a^2b = a(ab) = 0$ and $b \neq 0$. Next assume that $a^2b = 0$ for some $b \neq 0$. If $ab = 0$, then a is a zero-divisor. If $ab \neq 0$, then $a^2b = a(ab) = 0$ and we are done.

9. Take $a = (1,1,0)$, $b = (1,0,1)$ and $c = (0,1,1)$.

10. $\{(2,b,c) \mid b \in Z_5, c \in Z_6\} \cup \{(a,b,c) \mid a \in Z_4, b \in Z_5, c = 2,3,4\}$.

11. $(a_1 + b_1\sqrt{d}) - (a_2 + b_2\sqrt{d}) = (a_1 - a_2) + (b_1 - b_2)\sqrt{d}$;
$(a_1 + b_1\sqrt{d})(a_2 + b_2\sqrt{d}) = (a_1a_2 + b_1b_2d) + (a_1b_2 + a_2b_1)\sqrt{d}$. Thus the set is a ring. Since $Z[\sqrt{d}]$ is a subring of the ring of complex numbers, it has no zero-divisors.

12. Let $\frac{1}{2} = x$. Then $2x = 1$. So, $x = 4$. Let $-\frac{2}{3} = x$. Then $-2 = 3x$ which means $5 = 3x$. So $x = 4$. Note that $\sqrt{-3} = \sqrt{4} = 2$ or 5. $-\frac{1}{6} = \frac{-1}{6} = \frac{6}{6} = 1$.

13. The ring of even integers does not have a unity.

14. Look in Z_6.

15. Suppose that $a^2 = a$ for some a in Z_{p^n}. Then $a^2 - a = 0$ and therefore, p^n divides $a^2 - a = a(a-1)$. So, by Euclid's Lemma, p divides a or p divides $a - 1$. But p cannot divide both a and $a - 1$, for otherwise, p would divide $a - (a - 1) = 1$. Thus, p^n divides a or p^n divides $a - 1$. This means that $a = 0$ or $a - 1 = 0$.

16. First note that $a^2 = a$ implies that $a^3 = a^2 = a$. Then
$(a-1)^3 = a^3 - 3a^2 + 3a - 1 = a - 1$.

17. $a^2 = a$ implies $a(a-1) = 0$.

18. Suppose $a \neq 0$ and $a^n = 0$, where we take n to be as small as possible. Then $a \cdot 0 = 0 = a^n = a \cdot a^{n-1}$, so by cancellation, $a^{n-1} = 0$. This contradicts the assumption that n was as small as possible.

19. If $a^2 = a$ and $b^2 = b$, then $(ab)^2 = a^2b^2 = ab$. The other cases are similar.

20. Note that if n can be written is the form p^2m where p is a prime, then $(pm)^2 = p^2m^2 = nm = 0$ in Z_n and pm is not 0. On the other hand, if n is of the form $p_1p_2\cdots p_t$ where the p_i are distinct primes and $a^k = 0 \bmod n$ it follows from Euclid's Lemma that each p_i divides a. Thus $a = 0$ in Z_n.

21. Let $f(x) = x$ on $[-1,0]$ and $f(x) = 0$ on (0,1] and $g(x) = 0$ on $[-1,0]$ and $g(x) = x$ on (0,1]. Then $f(x)$ and $g(x)$ are in R and $f(x)g(x) = 0$ on $[-1,1]$.

22. We proceed by induction. The $n = 1$ case is trivial. Assume that $a^n = a$. Then $a^{n+1} = aa^n = aa = a$.

23. Suppose that a is an idempotent and $a^n = 0$. By the previous exercise, $a = 0$.

24. $(3+4i)^2 = 3+4i$; $(3+i)^2 = 3+i$.

25. There are four in all. Since $|i| = |2i| = 4$, all we need do is use the table to find an element whose square is i or $2i$. These are $1+i, 1+2i, 2+i$, and $2+2i$.

26. Units: $(1,1), (1,5), (2,1), (2,5)$;
zero-divisors: $\{(a,b) \mid a \in \{0,1,2\}, b \in \{2,3,4\}\}$;
idempotents: $\{(a,b) \mid a = 0,1,\ b = 1,3,4\}$;
nilpotents: $(0,0)$.

27. $a^2 = a$ implies $a(a-1) = 0$. So if a is a unit, $a - 1 = 0$ and $a = 1$.

28. **a.** If $f(a) = 0$ for some a, let $g(x) = 0$ for all $x \neq a$ and $g(a) = 1$. Then $f(x)(g(x) = 0$.
b. $f(x) = 0$
c. If $f(x)$ is never 0, then $1/f(x)$ is defined for all x.

29. Since F is commutative so is K. The assumptions about K satisfy the conditions for the One-Step Subgroup Test for addition and for multiplication (excluding the 0 element). So, K is a subgroup under addition and a subgroup under multiplication (excluding 0). Thus K is a subring in which every nonzero element is a unit.

30. The proof that $Q[\sqrt{d}]$ is a subring is the same as in Exercise 11. Moreover, if a and b are not both 0, then $(a + b\sqrt{d})^{-1} = a/(a^2 - db^2) - (b/(a^2 - db^2)\sqrt{d}$. This also shows that there are no zero divisors. To check that $a^2 - db^2 \neq 0$, note that if $\sqrt{d}$ is an integer then $Q[\sqrt{d}] = Q$. Otherwise, $a^2 - db^2 = 0$ implies that $\sqrt{d}$ is rational, which is false.

31. Note that $ab = 1$ implies $aba = a$. Thus $0 = aba - a = a(ba - 1)$. So, $ba - 1 = 0$.

32. 6 is the unity; 4 and 6 are their own inverses and 2 and 8 are inverses of each other. $\{0, 4, 8\}$

33. A subdomain of an integral domain D is a subset of D that is an integral domain under the operations of D. To show that P is a subdomain, note that $n \cdot 1 - m \cdot 1 = (n - m) \cdot 1$ and $(n \cdot 1)(m \cdot 1) = (mn) \cdot 1$ so that P is a subring of D. Moreover, $1 \in P$, P has no zero-divisors since D has none, and P is commutative because D is. Also, since every subdomain contains 1 and is closed under addition and subtraction, every subdomain contains P. Finally, we note that $|P| = \text{char } D$ when char D is prime and $|P|$ is infinite when char D is 0.

34. An integral domain of order 6 would be an Abelian group of order 6 under addition and, by Theorem 13.4, the characteristic is prime. But the only Abelian group of order 6 is Z_6, which by Theorem 13.3 has characteristic 6. This argument cannot be adapted to a ring with exactly four elements because the ring under addition could be isomorphic $Z_2 \oplus Z_2$, which has characteristic 2. The argument can be adapted for 15 elements. In general, there is no integral domain R with n elements when n is divisible by two distinct primes p and q because by the Fundamental Theorem of Finite Abelian Groups (or Theorem 9.5) R would have elements with additive orders p and q. So, $|1|$ would be a prime divisible by both p and q.

35. By Theorem 13.3, the characteristic is $|1|$ under addition. By Corollary 2 of Theorem 7.1, $|1|$ divides 2^n. By Theorem 13.4, the characteristic is prime. Thus, the characteristic is 2.

36. Solve the equation $x^2 = 1$.

37. By Exercise 36, 1 is the only element of an integral domain that is its own inverse if and only if $1 = -1$. This is true only for fields of characteristic 2.

38. If n is a prime, then Z_n is a field and therefore has no zero-divisors. If n is not a prime, we may write $n = ab$ where both a and b are less than n. If $a \neq b$, then $(n-1)!$ includes both a and b among its factors so $(n-1)! = 0$. If $a = b$ and $a > 2$, then $(n-1)! = (a^2-1)(a^2-2)\cdots(a^2-a)\cdots(a^2-2a)\cdots 2 \cdot 1$. Since this product includes $a^2 - a = a(a-1)$ and $a^2 - 2a = a(a-2)$, it contains $a^2 = n = 0$. The only remaining case is $n = 4$ and in this case $3! = 2$ is a zero-divisor.

39. **a.** First note that $a^3 = b^3$ implies that $a^6 = b^6$. Then $a = b$ because we can cancel a^5 from both sides (since $a^5 = b^5$).

b. Since m and n are relatively prime, by the corollary of Theorem 0.2, there are integers s and t such that $1 = sn + tm$. Since one of s and t, is negative we may assume that s is negative. Then
$a(a^n)^{-s} = a^{1-sn} = (a^m)^t = (b^m)^t = b^{1-sn} = b(b^n)^{-s} = b(a^n)^{-s}$. Now cancel $(a^n)^{-s}$.

Alternate proof for part **b.** Suppose there are relative prime positive integers m and n for which the statement is false. Among all such integers let $n < m$ and n be the minimum. Write $m = nq + r$ where $0 \leq r < n$. Since n does not divide m, $r \neq 0$. Then we can cancel $(a^n)^q$ on the right side of $a^m = b^m$ and $(b^n)^q$ on the left side to get $a^r = b^r$. Moreover, n and r are relative prime because any common divisor of r and n would be a common divisor of m and n. This contradicts the minimality of n.

40. In Z, take $a = 1$, $b = -1$, $m = 4$, $n = 2$.

41. If K is a subfield of F, then K^* is a subfield of F^*, which has order 31. So, $|K^*|$ must divide 31. This means that $|K^*| = 1$ or 31.

42.

	0	1	i	$1+i$
0	0	0	0	0
1	0	1	i	$1+i$
i	0	i	1	$1+i$
$1+i$	0	$1+i$	$1+i$	0

No. No.

43. In $Z_p[k]$, note that $(a + b\sqrt{k})^{-1} = \frac{1}{a+b\sqrt{k}} \frac{(a-b\sqrt{k})}{(a-b\sqrt{k})} = \frac{a-b\sqrt{k}}{a^2-b^2k}$ exists if and only if $a^2 - b^2k \neq 0$ where $a \neq 0$ and $b \neq 0$.

44. Observe that $(1 + i)^4 = -1$, so $|1 + i| = 8$ and therefore the group is isomorphic to Z_8.

45. Let a be a non-zero element. If the ring has n elements then the sequence $a, a^2, \ldots, a^{n+1}$ has two equal elements. Say $a^i = a^{k+i} = a^k a^i$. Let x be a non-zero element in the ring. Then $xa^k a^i = xa^i$ implies that $0 = xa^k a^i - xa^i = a^i(xa^k - x)$. So, $0 = xa^k - x$ and therefore $xa^k = x$.
Alternative solution: Let $S = \{a_1, a_2, \ldots, a_n\}$ be the nonzero elements of the ring. Then $a_1a_1, a_1a_2, \ldots, a_1a_n$ are distinct elements for if $a_1a_i = a_1a_j$, then $a_1(a_i - a_j) = 0$ and therefore $a_i = a_j$. If follows that $S = \{a_1a_1, a_1a_2, \ldots, a_1a_n\}$. Thus, $a_1 = a_1a_i$ for some i. Then a_i is the unity, for if a_k is any element of S, we have $a_1a_k = a_1a_ia_k$, so that $a_1(a_k - a_ia_k) = 0$. Thus, $a_k = a_ia_k$ for all k.

46. Say $(abb)c = 0$ where $c \neq 0$. If $ac = 0$, then a is a zero-divisor. If $ac \neq 0$, then $(ac)b = 0$ so that b is a zero-divisor.

47. Suppose that x and y are nonzero and $|x| = n$ and $|y| = m$ with $n < m$. Then $0 = (nx)y = x(ny)$. Since $x \neq 0$, we have $ny = 0$. This is a contradiction to the fact that $|y| = m$.

48. Use Exercise 47 and the observation that if $|a| = mn$, then $|ma| = n$.

49. **a.** For $n = 2$ the Binomial Theorem gives us
$(x_1 + x_2)^p = x_1^p + px_1^{p-1}x_2 + \cdots + px_1x_2^{p-1} + x_2^p$, where the coefficient $p!/k!(p - k)!$ of every term between x_1^p and x_2^p is divisible by p. Thus, $(x_1 + x_2)^p = x_1^p + x_2^p$. The general case follows by induction on n.

b. This case follows from Part a and induction on m.

c. Note Z_4 is a ring of characteristic 4 and $(1+1)^4 = 2^4 = 0$, but $1^4 + 1^4 = 1 + 1 = 2$.

50. Let the characteristic be p and $a^n = 0$. Then $p^n > n$ implies that $a^{p^n} = a^{p^n - n}a^n = 0$ and, by Exercise 49, $(1+a)^{p^n} = 1 + a^{p^n} = 1$.

51. By Theorem 13.4, $|1|$ has prime order, say p. Then, by Exercise 47, every nonzero element has order p. If the order of the field were divisible by a prime q other than p, Cauchy's Theorem (9.5) implies that the field also has an element of order q. Thus, the order of the field is p^n for some prime p and some positive integer n.

52. $Z_3[x]$

53. $n\begin{bmatrix} a & b \\ c & d \end{bmatrix} = \begin{bmatrix} 0 & 0 \\ 0 & 0 \end{bmatrix}$ for all members of $M_2(R)$ if and only if $na = 0$ for all a in R.

54. Observe char R = least common multiple $\{|x| \mid x \in R\}$ (additive order). Now use Corollary 2 to Theorem 7.1.

55. This follows directly from Exercise 54.

56. $2 + i$ and $2 + 2i$

57. **a.** 2 **b.** 2, 3 **c.** 2, 3, 6, 11 **d.** 2, 3, 9, 10

58. char S is a divisor of m. To verify, this let char $S = n$ and write $m = nq + r$ where $0 \leq r < n$. Then for all x in S we have $rx = (m - nq)x = mx - nqx = 0 - 0 = 0$. Since n is the least positive integer such that $nx = 0$ for all x in S we have $r = 0$. Alternate proof. Let char $S = n$. By Theorem 0.2 there are integers s and t such that $d = \gcd(m, n) = ms = nt$. Then for all x in S we have $dx = mxs + nxt = 0$. So, $d \geq n$. Since d is a divisor of n we have $n = d$.
Alternate proof. First observe that for a ring with positive characteristic, the characteristic is the least common multiple of the orders of the elements. Let char $S = n = p_1^{n_1}p_2^{n_2}\cdots p_k^{n_k}$ where the p_i are distinct primes. Then for each p_i there is an element s_i in S such that $p_i^{n_i}$ divides $|s_i|$. Thus $\langle s_i \rangle$ has an element s_i' of order p_i^{n-i}. Since $ms_i' = 0$, we have that $p_i^{n_i}$ divides m.

59. By Theorem 13.3, char R is prime. From $20 \cdot 1 = 0$ and $12 \cdot 1 = 0$ and Corollary 2 of Theorem 4.1, we know that char R divides both 12 and 20. Since the only prime that divides both 20 and 12 is 2, the characteristic is 2.

60. If $a^2 = a$ and $b^2 = b$, then $(a-b)^2 = a^2 + b^2 = a - b$ and $(ab)^2 = a^2b^2 = ab$.

61. Suppose $a \in Z_p$ and $a^2 + 1 = 0$. Then $(a+i)(a-i) = a^2 + 1 = 0$.

62. Let a be an idempotent other than 0 or 1. Then $a^2 = a$ implies that $a(a-1) = 0$.

63. Suppose that F is a field of order 16 and K is a subfield of F of order 8. Then K^* is a subgroup of F^* and $|K^*| = 7$ and $|F^*| = 15$, which contradicts Lagrange's theorem.

64. Use Corollary 4 of Theorem 7.1.

65. By Exercise 49, $x, y \in K$ implies that $x - y \in K$. Also, if $x, y \in K$ and $y \neq 0$, then $(xy^{-1})^p = x^p(y^{-1})^p = x^p(y^p)^{-1} = xy^{-1}$. So, by Exercise 29, K is a subfield.

66. Note that $(a+bi)-(c+di)=a-c+(b-d)\in Q(i)$. Also, in the complex numbers, $\frac{1}{a+bi}=\frac{1}{a+bi}\frac{a-bi}{a-bi}=\frac{a-bi}{a^2+b^2}=(a^2+b^2)^{-1}(a-bi)$, which is in $Q(i)$.

67. The unity is $(1,1,\ldots$. Let $S_n=\{0\}\oplus\{0\}\oplus\cdots\oplus Z_n\oplus\{0\}\oplus\{0\}\oplus\cdots$. Then S_n is a subring with characteristic n. This shows that R cannot have a positive characteristic.

68. $(1-a)^2=1-2a+a^2=1-2a+a=1-a$.

69. Observe that because $u_Su_R=u_S=u_Su_S$ we have $u_S(u_R-u_S)=0$.

70. If $\langle a\rangle=\langle b\rangle$, then there are elements u and v such that $a=bu$ and $b=va$. Then $a=bu=vau$ and $a(1-vu)=0$ and $vu=1$. If $a=bu$, then $a\in\langle b\rangle$ and if u is a unit then $b=au^{-1}\in\langle a\rangle$.

71. Since a field of order 27 has characteristic 3, we have $3a=0$ for all a. Thus, $6a=0$ and $5a=-a$.

72. Since the characteristic of a field of order 2^n is 2, it suffices to show that $a=b$ for then $0=a^2+ab+b^2=3a^2=2a^2+a^2=a^2$. Note that $a^3-b^3=(a-b)(a^2+ab+b^2)=0$ so that $a^3=b^3$. In a field of order 2^n, $x^{2^n+1}=x^2$ for all x (see Exercise 54). Also, for all odd n, 2^n+1 is divisible by 3. Thus $a^2=a^{2^n+1}=(a^3)^{(2^n+1)/3}=(b^3)^{(2^n+1)/3}=b^{2^n+1}=b^2$. Finally, $a^3=b^3$ and $a^2=b^2$ imply $a=b$.

73. Let $a\in F$, where $a\neq 0$ and $a\neq 1$. Then $(1+a)^3=1^3+3(1^2a)+3(1a^2)+a^3=1+a+a^2+a^3$. If $(1+a)^3=1^3+a^3$, then $a+a^2=0$. But then $a(1+a)=0$ so that $a=0$ or $a=-1=1$. This contradicts our choice of a.

74. $0,1,p,p+1$. To verify that this list is complete say that $a\in Z_{2p}$ and $a^2=a$, where $1<a<2p$. Then we have $a(a-1)=0$ in Z_{2p}. So, p divides a or p divides $a-1$. Thus, $a=p$ or $a-1=p$.

75. $0,1,p-1,p,p+1$. To verify that this list is complete say that $a\in Z_{2p}$ and $a^3=a$, where $1<a<2p$. Then we have $a(a-1)(a+1)=0$ in Z_{2p}. So, p divides a, or p divides $a-1$, or p divides $a+1$. Thus, $a=p$ or $a-1=p$ or $a+1=p$.

76. $Z_3\oplus Z_3\oplus\cdots\oplus Z_3$.

77. $Z_2\oplus Z_2\oplus\cdots\oplus Z_2$.

78. $f(x)=-x, f(x)=|x|, f(x)=\lfloor x\rfloor, f(x)=\lceil x\rceil$, and the function defined by $f(x)=1/x$ for $x\neq 0$ and $f(0)=0$.

79. Suppose that $Z_p[i]$ is a field and $p=1 \bmod 4$ and a and b are integers such that $a^2+b^2=p$. Then $(a+bi)(a-bi)=a^2+b^2=p=0$. So, $Z_p[i]$ has a zero-divisor. Now suppose that $p=3 \bmod 4$. By Theorem 13.2, it suffices to show that $Z_p[i]$ has no zero-divisors. If $a+bi$ is a zero-divisor, then so is $(a+bi)(a-bi)=a^2+b^2$ in $Z_p[x]$. But $(a^2+b^2)(x+yi)=0$ implies that $a^2+b^2=0 \bmod p$. So, there is an integer t such that $a^2+b^2=pt$ for some integer t. So, mod 4, we have $a^2+b^2=3t'$ for some t' in $\{0,1,2,3\}$. Trying all four cases for t' gives that mod 4, $a=b=0$ or $a=b=2$. So, a^2+b^2 is 0 or 4. If $a=b=0$, then $a+bi$ is not a zero-divisor. If $a=b=2$, then $a^2+b^2=4 \bmod p$, which is not 0 mod p.

80. Say that $ab=0$ and $b\neq 0$. Then, p^n divides ab. If p^n divides b, then $b=0$. So, p divides a. Say $a=pc$. Then, $a^{k-1}=p^{k-1}pc=p^kc=0$.

CHAPTER 14
Ideals and Factor Rings

1. Let r_1a and r_2a belong to $\langle a\rangle$. Then $r_1a - r_2a = (r_1 - r_2)a \in \langle a\rangle$. If $r \in R$ and $r_1a \in \langle a\rangle$, then $r(r_1a) = (rr_1)a \in \langle a\rangle$.

2. To prove that A is an ideal, note that $f(x) \in A$ if and only if $f(0) = 0$. If $f(x), g(x) \in A$, then $f(0) = 0$ and $g(0) = 0$. So, $f(0) - g(0) = 0 - 0 = 0$ and $h(0)f(0) = h(0)0 = 0$ for all $h(x)$ in $Z[x]$. Finally, note that $f(x) \in \langle x\rangle$ if and only if $f(0) = 0$.

3. Clearly, I is not empty. Now observe that $(r_1a_1 + \cdots + r_na_n) - (s_1a_1 + \cdots + s_na_n) = (r_1 - s_1)a_1 + \cdots + (r_n - s_n)a_n \in I$. Also, if $r \in R$, then $r(r_1a_1 + \cdots + r_na_n) = (rr_1)a_1 + \cdots + (rr_n)a_n \in I$. That $I \subseteq J$ follows from closure under addition and multiplication by elements from R.

4. $\{(a, a) \mid a \in Z\}$; $\{(a, -a) \mid a \in Z\}$.

5. Let $a + bi, c + di \in S$. Then $(a + bi) - (c + di) = a - c + (b - d)i$ and $b - d$ is even. Also, $(a + bi)(c + di) = ac - bd + (ad + cb)i$ and $ad + cb$ is even. Finally, $(1 + 2i)(1 + i) = -1 + 3i \notin S$.

6. **a.** $\langle 2\rangle$ **b.** $\langle 2\rangle$ and $\langle 5\rangle$ **c.** $\langle 2\rangle$ and $\langle 3\rangle$ **d.** $\langle p\rangle$ where p is a prime divisor of n.

7. Suppose that s is not prime. Then we can write $s = pm$ where p is prime and $m > 1$. Then $\langle s\rangle$ is properly contained in $\langle p\rangle$ and $\langle p\rangle$ is properly contained in Z_n. So $\langle s\rangle$ is not maximal. Now suppose that s is prime and there is a divisor $t > 1$ of n such that $\langle t\rangle$ properly contains $\langle s\rangle$ (recall every subgroup of Z_n has the form $\langle k\rangle$ where k is a divisor of n). Then $s = rt$ for some r. So we have $t = s$.

8. If $|a| = n$, then $n(ma) = m(na) = 0$.

9. If aR is an non-zero ideal of R, we know that $aR = R$. So, a belongs to R.

10. Because $2x \in \langle x, 3\rangle$ we have that $\langle 2x, 3\rangle \subseteq \langle x, 3\rangle$. Because $4x,\ 3x \in \langle 2x, 3\rangle$ we have $x \in \langle 2x, 3\rangle$. So, $\langle x, 3\rangle \subseteq \langle 2x, 3\rangle \subseteq$.

11. Since $ar_1 - ar_2 = a(r_1 - r_2)$ and $(ar_1)r = a(r_1r)$, aR is an ideal. $4R = \{\ldots, -16, -8, 0, 8, 16, \ldots\}$.

12. Mimic Exercise 9 of Chapter 12.

13. If n is a prime and $ab \in Z$, then by Euclid's Lemma (Chapter 0), n divides a or n divides b. Thus, $a \in nZ$ or $b \in nZ$. If n is not a prime, say $n = st$ where $s < n$ and $t < n$, then st belongs to nZ but s and t do not.

14. $(a_1 + b_1) - (a_2 + b_2) = (a_1 - a_2) + (b_1 - b_2) \in A + B$; $r(a + b) = ra + rb \in A + B$; $(a + b)r = ar + br \in A + B$

15. **a.** $a = 1$ **b.** $a = 2$ **c.** $a = \gcd(m, n)$

16. Let $a_1b_1 + \cdots + a_nb_n$ and $a'_1b'_1 + \cdots a'_mb'_m$ then

$$\begin{aligned}(a_1b_1 + \cdots + a_nb_b) - (a'_1b'_1 + \cdots + a'_mb'_m) \\ = a_1b_1 + \cdots + a_nb_n + (-a'_1)b'_1 + \cdots + (-a'_m)b'_m \in AB.\end{aligned}$$

Also

$$r(a_1b_1 + \cdots + a_nb_n) = (ra_1)b_1 + \cdots + (ra_n)b_n \in AB$$

and

$$(a_1b_1 + \cdots + a_nb_n)r = a_1(b_1r) + \cdots + a_n(b_nr) \in AB.$$

17. **a.** $a = 12$

b. $a = 48$. To see this, note that every element of $\langle 6\rangle\langle 8\rangle$ has the form $6t_18k_1 + 6t_28k_2 + \cdots + 6t_n8k_n = 48s \in \langle 48\rangle$. So, $\langle 6\rangle\langle 8\rangle \subseteq \langle 48\rangle$. Also, since $48 \in \langle 6\rangle\langle 8\rangle$, we have $\langle 48\rangle \subseteq \langle 6\rangle\langle 8\rangle$.

c. $a = mn$

18. Since A and B are ideals, $ab \in A$ and $ab \in B$ when $a \in A$ and $b \in B$. Now AB is just the sum of such terms.

19. Let $r \in R$. Then $r = 1r \in A$.

20. By Exercise 18, we have $AB \subseteq A \cap B$. So, let $x \in A \cap B$. To show that $x \in AB$, start by writing $1 = a + b$ where $a \in A, b \in B$.

21. Let $u \in I$ be a unit and let $r \in R$. Then $r = r(u^{-1}u) = (ru^{-1})u \in I$.

22. Let A be a prime ideal of R. By Theorem 14.3, R is an integral domain. Then, by Theorem 13.2, R/A is a field and, by Theorem 14.4, A is maximal.

23. Observe that $\langle 2\rangle$ and $\langle 3\rangle$ are the only nontrivial ideals of Z_6, so both are maximal. More generally, Z_{pq}, where p and q are distinct primes, has exactly two maximal ideals.

24. Observe that as groups $|R : I| = 3$. So there is not a proper subgroup of R that strictly contains I.

25. I is closed under subtraction since the even integers are closed under subtraction. Also, if b_1, b_2, b_3, and b_4 are even, then every entry of $\begin{bmatrix} a_1 & a_2 \\ a_3 & a_4 \end{bmatrix}\begin{bmatrix} b_1 & b_2 \\ b_3 & b_4 \end{bmatrix}$ is even.

26. $0 \in I$. Let $f(x), g(x) \in I$ and $h(x) \in Z[x]$. Then $f(1) - g(1)$ even and $f(1)h(1)$ is even.

27. The proof that I is an ideal is the same as the case that $n = 2$ in Exercise 25. The number of elements in I is n^4.

28. Since A is non-empty and closed under subtraction and multiplication, it is a subring. It is not an ideal because $\begin{bmatrix} 0 & 0 \\ 1 & 0 \end{bmatrix}\begin{bmatrix} 1 & 0 \\ 0 & 0 \end{bmatrix} = \begin{bmatrix} 0 & 0 \\ 1 & 0 \end{bmatrix}$ is not in A.

29. That I satisfies the ideal test follows directly from the definitions of matrix addition and multiplication. To see that R/I is a field, first observe that $\begin{bmatrix} a & b \\ 0 & c \end{bmatrix} + I = \begin{bmatrix} a & 0 \\ 0 & 0 \end{bmatrix} + \begin{bmatrix} 0 & b \\ 0 & c \end{bmatrix} + I = \begin{bmatrix} a & 0 \\ 0 & 0 \end{bmatrix} + I.$

Thus we need only show that $\begin{bmatrix} a & 0 \\ 0 & 0 \end{bmatrix} + I$ has an inverse in R/I when $a \neq 0$. To this end, note that

$$\left(\begin{bmatrix} a & 0 \\ 0 & 0 \end{bmatrix} + I\right)\left(\begin{bmatrix} a^{-1} & 0 \\ 0 & 0 \end{bmatrix} + I\right) = \begin{bmatrix} 1 & 0 \\ 0 & 0 \end{bmatrix} + I =$$

$$\left(\begin{bmatrix} 1 & 0 \\ 0 & 0 \end{bmatrix} + I\right) + \left(\begin{bmatrix} 0 & 0 \\ 0 & 1 \end{bmatrix} + I\right) = \begin{bmatrix} 1 & 0 \\ 0 & 1 \end{bmatrix} + I.$$

$$\left(\begin{bmatrix} 2 & 0 \\ 0 & 0 \end{bmatrix} + I\right)^{-1} = \begin{bmatrix} 1/2 & 0 \\ 0 & 0 \end{bmatrix} + I.$$

30. Since $(-i)i = -i^2 = 1 \in \langle i \rangle$ every element of $Z[i]$ is in $\langle i \rangle$. So, $Z[x]/\langle i \rangle = \{0 + \langle i \rangle\}$.

31. $R = \{0 + \langle 2i \rangle, 1 + \langle 2i \rangle, i + \langle 2i \rangle, 1 + i + \langle 2i \rangle\}$. R is not an integral domain because $(1 + i + \langle 2i \rangle)^2 = (1+i)^2 + \langle 2i \rangle = 1 + 1 + \langle 2i \rangle = 0 + \langle 2i \rangle$.

32. $1 = |1| < |2 - i||a + bi| = \sqrt{5}\sqrt{a^2 + b^2}$.

33. Let $I = \langle x^2 - 2x - 3, x \rangle$. Observe that $x \in I$ implies that x^2 and $2x$ belong to I. Then, so does $(x^2 - 2x - 3) + (x^2 + 2x) = -3$. But then $(-3)(-2) = 1$ belongs to I. So, $I = Z_5[x]$.

34. First note that $I[x]$ is the set of all polynomials in $Z[x]$ for which every coefficient is even. Since the element $3 + I[x]$ does not have a multiplicative inverse in $Z[x]/I[x]$, it not a field. So, by Theorem 14.4, $I[x]$ is not maximal.

35. Use the observation that every member of R can be written in the form $\begin{bmatrix} 2q_1 + r_1 & 2q_2 + r_2 \\ 2q_3 + r_3 & 2q_4 + r_4 \end{bmatrix}$ where each r_i is 0 or 1. Then note that

$$\begin{bmatrix} 2q_1 + r_1 & 2q_2 + r_2 \\ 2q_3 + r_3 & 2q_4 + r_4 \end{bmatrix} + I = \begin{bmatrix} r_1 & r_2 \\ r_3 & r_4 \end{bmatrix} + I.$$

36. Let R be the ring $\{0, 2, 4, 6\}$ under addition and multiplication mod 8. Then $\{0, 4\}$ is maximal but not prime.

37. $(br_1 + a_1) - (br_2 + a_2) = b(r_1 - r_2) + (a_1 - a_2) \in B$; $r'(br + a) = b(r'r) + r'a \in B$ since $r'a \in A$.

38. $A = \langle 2 \rangle$.

39. Suppose that I is an ideal of F and $I \neq \{0\}$. Let a be a nonzero element of I. Then by Exercise 21, $I = F$.

40. Since for every nonzero element a in R, aR is a nonzero ideal of R, we have $R = aR$. Thus, there is an element b in R such $1 = ab$. This shows that every non-zero element in R has an inverse in R. So, R is a field.
Alternate proof. Since the only ideal that properly contains $\{0\}$ is R, $\{0\}$ is a maximal ideal of R. So, by Theorem 14.4 $R/0$ is a field. So, for any non-zero element a in R there is an element b in R such that $1 + \{0\} = (a + \{0\})(b + \{0\}) = ab + \{0\}$. Thus $\{1\} = \{ab\}$ and $1 = ab$.

41. Let a be an idempotent other that 0 or 1. Then $a^2 = a$ implies that $a(a - 1) = 0$.

42. Use Example 15 and Theorem 14.4.

43. Since every element of $\langle x\rangle$ has the form $xg(x)$, we have $\langle x\rangle \subseteq I$. If $f(x) \in I$, then $f(x) = a_nx^n + \cdots + a_1x = x(a_nx^{n-1} + \cdots + a_1) \in \langle x\rangle$. We know that I is not maximal because $J = \{f(x) \in Z[x] \mid f(0) \text{ is even}\}$ is a proper ideal of $Z[x]$ that properly contains I.

44. Since $A = \langle(3,0),(0,1)\rangle$, it is an ideal. Then observe that $Z \oplus Z/A = \{(0,0)+A,(1,0)+A,(2,0)+A\} \approx Z_3$ and use Theorem 14.4. In general, for $A = \{(nx,y) \mid x,y \in Z\}$, $Z \oplus Z/A =$ $\{(0,0)+A,(1,0)+A,(2,0)+A,\ldots,(n-1,0)+A\} \approx Z_n$. Thus, A is a maximal ideal of $Z \oplus Z$ if and only if n is prime.

45. Suppose J is an ideal that properly contains I and let $f(x) \in J$ but not in I. Then, since $g(x) = f(0) - f(x)$ belongs to I and $f(x)$ belongs to J, we know that $f(0) = g(x) + f(x)$ is a non-zero constant contained in J. So, by Exercise 17, $J = R$.

46. $\langle 1\rangle \oplus \langle 2\rangle$, $\langle 2\rangle \oplus \langle 1\rangle$, $\langle 1\rangle \oplus \langle 3\rangle$, $\langle 1\rangle \oplus \langle 5\rangle$; 2, 2, 3, 5.

47. Since $(3+i)(3-i) = 10$ we know $10 + \langle 3+i\rangle = 0 + \langle 3+i\rangle$. Also, $i + \langle 3+i\rangle = -3 + \langle 3+i\rangle = 7 + \langle 3+i\rangle$. Thus, every element $a + bi + \langle 3+i\rangle$ can be written in the form $k + \langle 3+i\rangle$ where $k = 0,1,\ldots,9$. Finally, $Z[i]/\langle 3+i\rangle = \{k + \langle 3+i\rangle \mid k = 0,1,\ldots,9\}$, since $1 + \langle 3+i\rangle$ has additive that divides order 10. To see that the additive order is not 5, we note that if so then there are integers a and b such that $5 = (3+i)(a+bi) = 3a - b + (a+3b)i$. This gives us the two equations $5 = 3a - b$ and $0 = a + 3b$. But these have no solution in Z. The same argument rules out the orders 1 and 2.

48. Since $I = \{2m + 2n)i \mid m,n \in Z\} = \langle 2\rangle$, it is an ideal. Elements are $\{0+i, 1+I, i+I, 1+i+I\}$. Note that $(1+i+I)^2 = 0+I$, so $Z[i]/I$ is not a field or an integral domain.

49. Because $I = \langle(1,0)\rangle$, I is an ideal. To prove that I is prime suppose that $(a,0)(b,0) = (ab,0) = (0,0)$. Then $ab = 0$ and therefore $a = 0$ or $b = 0$. So, I is prime. Finally, because $(2,0)$ has no multiplicative inverse in $Z \oplus Z$, $A \oplus Z)/I$ is not a field and I is not maximal. Or note that I is a proper subring of the ideal $J = \{(a,b) \mid a,b \in Z, \text{ and } b \text{ is even}\}$.

50. $rs - sr \in I$ if and only if $rs - sr + I = I$ or $rs + I = sr + I$. This is equivalent to $(r+I)(s+I) = (s+I)(r+I)$.

51. Since every element in $\langle x,2\rangle$ has the form $f(x) = xg(x) + 2h(x)$, we have $f(0) = 2h(0)$, so that $f(x) \in I$. If $f(x) \in I$, then $f(x) = a_nx^n + \cdots + a_1x + 2k = x(a_nx^{n-1} + \cdots + a_1) + 2k \in \langle x,2\rangle$. By Theorems 14.3 and 14.4, to prove that I is prime and maximal, it suffices to show that $Z[x]/I$ is a field. To this end, note that every element of $Z[x]/I$ can be written in the form $a_nx^n + \cdots + a_1x + 2k + I = 0 + I$ or $a_nx^n + \cdots + a_1x + (2k+1) + I = 1 + I$. So, $Z[x]/I \approx Z_2$.

52. $I = \langle 2x, 4\rangle$.

53. One example is $J = \langle x^2+1, 2\rangle$. To see that 1 is not in J, note that if there were $f(x), g(x) \in Z[x]$ such that $(x^2+1)f(x) + 2g(x) = 1$, then evaluating the left side at 1 yields an even integer.

54. We first determine the characteristic of $Z[i]/I$. Note that $2(1+i)(1-i) = 2(1+1) = 4$ is in $\langle 2+2i\rangle$ and therefore

$4(1 + \langle 2+2i \rangle) = 0 + \langle 2+2i \rangle$. So, the characteristic is 4 or 2. But $2(1 + \langle 2+2i \rangle) \neq 0 + \langle 2+2i \rangle$. So, by Theorem 13.3, the characteristic of $Z[i]/I$ is 4. Because $0 + I = 2 + 2i + I$, we have that $2i + I = -2 + I = 2 + I$. So, every element of $Z[i]/I$ can be written in the form $a + I$ or $a + i + I$ where $a = 0, 1, 2$, or 3. Finally, as in Example 11, we can show that i is not in I so these 8 elements are distinct. Finally, because $2 \cdot 2 \in I$ and 2 is not, I is not prime.

55. $3x + 1 + I$

56. Let $a, b \in I_p$. Say $|a| = p^n$ and $|b| = p^m$. Then $p^{n+m}(a-b) = 0$ so $|a-b|$ divides p^{n+m}. Also, $p^n(ra) = r(p^n a) = 0$ so $|ra|$ divides p^n.

57. Every ideal is a subgroup. Every subgroup of a cyclic group is cyclic.

58. Since $\begin{bmatrix} a & b \\ 0 & d \end{bmatrix}\begin{bmatrix} r & s \\ 0 & t \end{bmatrix} = \begin{bmatrix} ar & as+bt \\ 0 & dt \end{bmatrix}$ and $\begin{bmatrix} r & s \\ 0 & t \end{bmatrix}\begin{bmatrix} a & b \\ 0 & d \end{bmatrix} = \begin{bmatrix} ra & rb+sd \\ 0 & td \end{bmatrix}$ for all a, b, and d, we must have that r and t are even.

59. Let I be any ideal of $R \oplus S$ and let $I_R = \{r \in R |\ (r,s) \in I \text{ for some } s \in S\}$ and $I_S = \{s \in S |\ (r,s) \in I \text{ for some } r \in R\}$. Then I_R is an ideal of R and I_S is an ideal of S. Let $I_R = \langle r \rangle$ and $I_S = \langle s \rangle$. Since, for any $(a,b) \in I$ there are elements $a' \in R$ and $b' \in S$ such that $(a,b) = (a'r, b's) = (a', b')(r,s)$, we have that $I = \langle (r,s) \rangle$.

60. Suppose A is prime and B is an ideal which properly contains A. Say $A = \langle a \rangle$ and $B = \langle b \rangle$. It suffices to show b is a unit. Write $a = br$. Then, since A is prime, $b \in A$ or $r \in A$. If $b \in A$, then $B \subseteq A$ so $r \in A$. Say $r = ar'$. Then $a = br = bar'$ so that $a(1 - br') = 0$. Thus $1 = br'$ and b is a unit.

61. Say $b, c \in \text{Ann}(A)$. Then $(b-c)a = ba - ca = 0 - 0 = 0$. Also, $(rb)a = r(ba) = r \cdot 0 = 0$.

62. Suppose $b, c \in N(A)$. Say, $b^n \in A$ and $c^m \in A$. Then the binomial theorem shows that $(b-c)^{n+m} \in A$. Also, $(rb)^n = r^n b^n \in A$.

63. **a.** $\langle 3 \rangle$ **b.** $\langle 3 \rangle$ **c.** $\langle 3 \rangle$

64. **a.** $\langle 6 \rangle$ **b.** $\langle 2 \rangle$ **c.** $\langle 6 \rangle$

65. Suppose $(x + N(\langle 0 \rangle))^n = 0 + N(\langle 0 \rangle)$. We must show that $x \in N(\langle 0 \rangle)$. We know that $x^n + N(\langle 0 \rangle) = 0 + N(\langle 0 \rangle)$, so that $x^n \in N(\langle 0 \rangle)$. Then, for some m, $(x^n)^m = 0$, and therefore $x \in N(\langle 0 \rangle)$.

66. Clearly $N(A) \subseteq N(N(A))$. Suppose $x \in N(N(A))$. Then $x^n \in N(A)$ for some n. Thus $(x^n)^m \in A$ for some m, so $x \in N(A))$.

67. Let $I = \langle x^2 + x + 1 \rangle$. Then $Z_2[x]/I = \{0 + I, 1 + I, x + I, x + 1 + I\}$. $1 + I$ is its own multiplicative inverse and $(x + I)(x + 1 + I) = x^2 + x + I = x^2 + x + 1 + 1 + I = 1 + I$. So, every nonzero element of $Z_2[x]/I$ has a multiplicative inverse.

68. $R = \{0 + \langle 2i \rangle, 1 + \langle 2i \rangle, i + \langle 2i \rangle, 1 + i + \langle 2i \rangle\}$ R is not an integral domain because $(1 + i + \langle 2i \rangle)^2 = (1+i)^2 + \langle 2i \rangle = 1 + 1 + \langle 2i \rangle = 0 + \langle 2i \rangle$.

69. $x + 2 + \langle x^2 + x + 1 \rangle$ is not zero, but its square is.

70. $\{na + ba \mid n \in Z, b \in R\}$

71. If f and $g \in A$, then $(f-g)(0) = f(0) - g(0)$ is even and $(f \cdot g)(0) = f(0) \cdot g(0)$ is even. $f(x) = 1/2 \in R$ and $g(x) = 2 \in A$, but $f(x)g(x) \notin A$.

72. Observe that $1 + \langle 1-i \rangle = i + \langle 1-i \rangle$ so any coset can be written in the form $a + \langle 1-i \rangle$ where $a \in Z$. But

$$\begin{aligned} 1 + \langle 1-i \rangle &= (1 + \langle 1-i \rangle)^2 \\ &= (i + \langle 1-i \rangle)^2 = -1 + \langle 1-i \rangle \end{aligned}$$

so $2 + \langle 1-i \rangle = 0 + \langle 1-i \rangle$. This means that there are only two cosets: $0 + \langle 1-i \rangle$ and $1 + \langle 1-i \rangle$.

73. Any ideal of R/I has the form A/I where A is an ideal of R. So, if $A = \langle a \rangle$, then $A/I = \langle a + I \rangle / I$.

74. There is 1 element. To see this, let $I = \langle 1+i \rangle$. Then $(1+i)(1-i) = 2$ belongs to I and therefore $3 \cdot 2 = 1$ is in I. It follows that $I = Z_5[i]$ and the only element in the factor ring is $0 + I$.

75. In Z, $\langle 2 \rangle \cap \langle 3 \rangle = \langle 6 \rangle$ is not prime.

76. Suppose that (a, b) is a nonzero element of an ideal I in $\mathbf{R} \oplus \mathbf{R}$. If $a \neq 0$, then $(r, 0) = (ra^{-1}, 0)(a, b) \in I$. Thus, $\mathbf{R} \oplus \{0\} \subseteq I$. Similarly, if $b \neq 0$, then $\{0\} \oplus \mathbf{R} \subseteq I$. So, the ideals of $\mathbf{R} \oplus \mathbf{R}$ are $\{0\} \oplus \{0\}, \mathbf{R} \oplus \mathbf{R}, \mathbf{R} \oplus \{0\}, \{0\} \oplus \mathbf{R}$. The ideals of $F \oplus F$ are $\{0\} \oplus \{0\}, F \oplus F, F \oplus \{0\}, \{0\} \oplus F$.

77. According to Theorem 13.3, we need only determine the additive order of $1 + \langle 2+i \rangle$. Since $5(1 + \langle 2+i \rangle) = 5 + \langle 2+i \rangle = (2+i)(2-i) + \langle 2+i \rangle = 0 + \langle 2+i \rangle$, we know that $1 + \langle 2+i \rangle$ has order 5.

78. Use the fact that $a^2 + b^2 = (a+bi)(a-bi)$.

79. The set K of all polynomials whose coefficients are even is closed under subtraction and multiplication by elements from $Z[x]$ and therefore K is an ideal. By Theorem 14.3, to show that K is prime, it suffices to show that $Z[x]/K$ has no zero-divisors. Suppose that $f(x) + K$ and $g(x) + K$ are nonzero elements of $Z[x]/K$. Since K absorbs all terms that have even coefficients, we may assume that $f(x) = a_m x^m + \cdots + a_0$ and $g(x) = b_n x^n + \cdots + b_0$ are in $Z[x]$ and a_m and b_n are odd integers. Then $(f(x) + K)(g(x) + K) = a_m b_n x^{m+n} + \cdots + a_0 b_0 + K$ and $a_m b_n$ is odd. So, $f(x)g(x) + K$ is nonzero.

80. That I is closed under subtraction follows from the definition of addition in the complex number. So, let $a + b\sqrt{-5} \in I$ and let $c + d\sqrt{-5} \in R$. Then $(a + b\sqrt{-5})(c + d\sqrt{-5}) = (ac - 5bd) + (bc + ad)\sqrt{-5}$. So we must show that $(ac - 5bd) - (bc + ad) = (a-b)c + (-5b - a)d$ is even when $a - b$ is even. This reduces to showing that $-5b - a$ is even. Note that $a - b$ is even means that both a and b are odd or both a and b are even. Either way, $-5b - a$ is even.

81. By Theorem 14.3, R/I is an integral domain. Since every element in R/I is an idempotent and Exercise 16 in Chapter 13 says that the only idempotents in an integral domain are 0 and 1, we have that $R/I = \{0 + I, 1 + I\}$.

82. Say $J \neq I$ is also a maximal ideal. Let x be an element of J that is not in I. Then $1 = x^{-1}x \in J$ so that $J = R$.

83. $\langle x \rangle \subset \langle x, 2^n \rangle \subset \langle x, 2^{n-1} \rangle \subset \cdots \subset \langle x, 2 \rangle$

84. Use the ideal test to show that I is an ideal of R. To show that I is not generated by a single element observe that every element of I has the form $(b_1, b_2, b_3, \ldots, b_k, 0, 0, 0, \ldots)$ where all terms

85. Suppose that J is an ideal that properly contains I. Then J/I is a subring of R/I with more than one element and $|J/I|$ divides $|R/I|$. So $J = R$.

86. Let $R = Z_4$ and $I = \langle 2 \rangle$.

87. Look at coset representatives and use the fact that $(a + bx)^{-1} = \frac{1}{ax+b} = \frac{1}{ax+b}\frac{ax-b}{ax-b} = \frac{ax-b}{-a^2-b^2}$. Then check to see if this works.

CHAPTER 15

Ring Homomorphisms

1. Property 1: $\phi(nr) = n\phi(r)$ holds because a ring is a group under addition. To prove that $\phi(r^n) = (\phi(r))^n$, we note that by induction
$\phi(r^n) = \phi(r^{n-1}r) = \phi(r^{n-1})\phi(r) = \phi(r)^{n-1}\phi(r) = \phi(r)^n$.
Property 2: If $\phi(a)$ and $\phi(b)$ belong to $\phi(A)$, then $\phi(a) - \phi(b) = \phi(a - b)$ and $\phi(a)\phi(b) = \phi(ab)$ belong to $\phi(A)$.
Property 3: $\phi(A)$ is a subgroup because ϕ is a group homomorphism. Let $s \in S$ and $\phi(r) = s$. Then $s\phi(a) = \phi(r)\phi(a) = \phi(ra)$ and $\phi(a)s = \phi(a)\phi(r) = \phi(ar)$.
Property 4: Let a and b belong to $\phi^{-1}(B)$ and r belong to R. Then $\phi(a)$ and $\phi(b)$ are in B. So, $\phi(a) - \phi(b) = \phi(a) + \phi(-b) = \phi(a - b) \in B$. Thus, $a - b \in \phi^{-1}(B)$. Also, $\phi(ra) = \phi(r)\phi(a) \in B$ and $\phi(ar) = \phi(a)\phi(r) \in B$. So, ra and $ar \in \phi^{-1}(B)$.
Property 5: $\phi(a)\phi(b) = \phi(ab) = \phi(ba) = \phi(b)\phi(a)$.
Property 6: Because ϕ is onto, every element of S has the form $\phi(a)$ for some a in R. Then $\phi(1)\phi(a) = \phi(1a) = \phi(a)$ and $\phi(a)\phi(1) = \phi(a1) = \phi(a)$.
Property 7: If ϕ is an isomorphism, by property 1 of Theorem 10.1 and the fact that ϕ is one-to-one, we have Ker $\phi = \{0\}$. If Ker $\phi = \{0\}$, by property 5 of Theorem 10.2, ϕ is one-to-one.
Property 8: That ϕ^{-1} is one-to-one and preserves addition comes from property 3 of Theorem 6.3. To see that ϕ^{-1} preserves multiplication, note that
$\phi^{-1}(ab) = \phi^{-1}(a)\phi^{-1}(b)$ if and only if
$\phi(\phi^{-1}(ab)) = \phi(\phi^{-1}(a)\phi^{-1}(b)) = \phi(\phi^{-1}(a))\phi(\phi^{-1}(b))$. But this reduces to $ab = ab$.

2. Since ϕ is a group homomorphism, Ker ϕ is a subgroup. Let $a \in$ Ker ϕ and $r \in R$. The $\phi(ar) = \phi(a)\phi(r) = 0\phi(r) = 0$. Similarly, $\phi(ra) = 0$.

3. We already know the mapping is an isomorphism of groups. Let $\Phi(x + \text{Ker}\,\phi) = \phi(x)$. Note that $\Phi((r + \text{Ker}\,\phi)(s + \text{Ker}\,\phi)) = \Phi(rs + \text{Ker}\,\phi) = \phi(rs) = \phi(r)\phi(s) = \Phi(r + \text{Ker}\,\phi)\Phi(s + \text{Ker}\,\phi)$.

4. See the proof of Theorem 10.4.

5. $\phi(2 + 4) = \phi(1) = 5$, whereas $\phi(2) + \phi(4) = 0 + 0 = 0$.

6. $xy \to 3xy \neq 3x3y$.

7. Observe that $(x + y)/1 = (x/1) + (y/1)$ and $(xy)/1 = (x/1)(y/1)$.

8. If ϕ is a ring-homomorphism from Z_n to itself then $\phi(x) = \phi(1x) = \phi(1)x$. Moreover, if $\phi(1) = a$, then $a^2 = (\phi(1))^2 = \phi(1^2) = \phi(1) = a$.

9. The condition that the order of k divides both m and n is necessary and sufficient for the mapping to be group homomorphism is Exercise 5 in Chapter 10.
So, we need to show that k is an idempotent in Z_n is necessary and sufficient for $\phi(ab) = \phi(a)\phi(b)$. If $\phi(ab) = \phi(a)\phi(b)$, then, in Z_n, $k = \phi(1) = \phi(1 \cdot 1) = \phi(1)\phi(1) = k^2$. On the other hand, if $k^2 = k$ in Z_n, then $\phi(ab) = k(ab) = k^2(ab) = (ka)(kb) = \phi(a)\phi(b)$.

10. Let ϕ be the mapping that takes $\left\{\begin{bmatrix} a_1 & a_2 \\ a_3 & a_4 \end{bmatrix} \middle| \, a_i \in Z\right\}$ to $\left\{\begin{bmatrix} a_1 \bmod 2 & a_2 \bmod 2 \\ a_3 \bmod 2 & a_4 \bmod 2 \end{bmatrix}\right\}$ and use Theorem 15.3.

11. For groups, $\phi(x) = ax$ for $a = 2, 4, 6, 8$ since each of these has additive order 5. For rings, only $\phi(x) = 6x$ since 6 is the only non-zero idempotent in R.

12. Parts **a** and **b**. No. Suppose $2 \to a$. Then $4 = 2 + 2 \to a + a = 2a$ and $4 = 2 \cdot 2 \to aa = 2a$.

13. If a and $b\,(b \neq 0)$ belong to every member of the collection, then so do $a - b$ and ab^{-1}. Thus, by Exercise 29 of Chapter 13, the intersection is a subfield.

14. Try $a + bi \to a + bx + \langle x^2 + 1\rangle$.

15. By observation, ϕ is one-to-one and onto. Since
$\phi((a+bi)+(c+di)) = \phi((a+c)+(b+d)i) = \begin{bmatrix} a+c & b+d \\ -(b+d) & a+c \end{bmatrix} =$
$\begin{bmatrix} a & b \\ -b & a \end{bmatrix} + \begin{bmatrix} c & d \\ -d & c \end{bmatrix} = \phi(a+bi) + \phi(c+di)$

addition is preserved. Also,

$\phi((a+bi)(c+di)) = \phi((ac-bd)+(ad+bc)i) = \begin{bmatrix} ac-bd & ad+bc \\ -(ad+bc) & ac-bd \end{bmatrix} =$

$\begin{bmatrix} a & b \\ -b & a \end{bmatrix}\begin{bmatrix} c & d \\ -d & c \end{bmatrix} = \phi(a+bi)\phi(c+di)$

so multiplication is preserved.

16. Try $a + b\sqrt{2} \to \begin{bmatrix} a & 2b \\ b & a \end{bmatrix}$.

17. Since $\phi\left(\begin{bmatrix} a & b \\ c & d \end{bmatrix}\begin{bmatrix} a' & b' \\ c' & d' \end{bmatrix}\right) = \phi\left(\begin{bmatrix} aa'+bc' & ab'+bd' \\ ca'+dc' & cb'+dd' \end{bmatrix}\right) = aa' + bc' \neq aa' =$
$\phi\left(\begin{bmatrix} a & b \\ c & d \end{bmatrix}\right)\phi\left(\begin{bmatrix} a' & b' \\ c' & d' \end{bmatrix}\right)$, multiplication is not preserved.

18. It is a ring homomorphism.

19. Yes. $\phi(x) = 6x$ is well defined because $a = b$ in Z_5 implies that 5 divides $a - b$. So, 30 divides $6a - 6b$. Moreover, $\phi(a+b) = 6(a+b) = 6a + 6b = \phi(a) + \phi(b)$ and $\phi(ab) = 6ab = 6 \cdot 6ab = 6a6b = \phi(a)\phi(b)$.

20. No. For $\phi(x) = 2x$ we have $2 = \phi(1) = \phi(1 \cdot 1) = \phi(1)\phi(1) = 4$.

21. The set of all polynomials passing through the point $(1, 0)$.

22. Z

23. If ϕ is a ring homomorphism from Z_n to Z_{p^2}, then $|\phi(1)|$ is 1 or p, since it must divide both n and p^2. So, $\phi(1)$ is in $\langle p\rangle$. Because the only idempotent in $\langle p\rangle$ is 0, $\phi(x) = 0$.

24. Say a ring homomorphism ϕ maps 1 to a. Then the additive order of a must divide 25 and 20. So $|a| = 1$ or 5 and therefore $a = 0, 4, 8, 12$ or 16. But $1 = 1^1$ means that $a = a^2$. Checking each possibility we obtain that $a = 0$ or 16. Both of those give ring homomorphisms.

25. For Z_6 to Z_6, $1 \to 0$, $1 \to 1$, $1 \to 3$, and $1 \to 4$ each define a homomorphism. For Z_{20} to Z_{30}, $1 \to 0$, $1 \to 6$, $1 \to 15$, and $1 \to 21$ each define a homomorphism.

26. By property 6 of Theorem 15.1, 1 must map to 1. Thus, the only ring-isomorphism of Z_n to itself is the identity.

27. Suppose that ϕ is a ring homomorphism from Z to Z. Let $\phi(1) = a$. By property 1 of Theorem 15.1, $\phi(m) = \phi(m \cdot 1) = m\phi(1) = ma$. Moreover, $a = \phi(1) = \phi(1 \cdot 1) = (\phi(1))^2 = a^2$. This gives us that $a = 0$ or 1. So, ϕ is the zero map or $\phi(m) = m$ for all integers m.

28. Since $(1, 0)$ is an idempotent and idempotents must map to idempotents; the possibilities are $(0, 0), (1, 0), (0, 1), (1, 1)$.

29. If $ab = 1$, then $\phi(a)\phi(b) = \phi(1)$, which is a unity in S (see property 7 of Theorem 15.1). If a is a zero-divisor in R, then $\phi(a)$ is a zero-divisor in S or is 0.

30. The group A/B is cyclic of order 4. The ring A/B has no unity.

31. First, note that every element of $R[x]/\langle x^2 \rangle$ can be written uniquely in the form $ax + a + \langle x^2 \rangle$. Then mapping that takes $ax + b + \langle x^2 \rangle$ to $\left\{ \begin{bmatrix} a & b \\ 0 & a \end{bmatrix} \right\}$ is a ring isomorphism.
Alternate proof: The mapping $a_n x^n + a_{n-1}x^{n-1} + \cdots + a_1 x + a_0$ to $\left\{ \begin{bmatrix} a_0 & a_1 \\ 0 & a_0 \end{bmatrix} \right\}$ is a ring homomorphism with kernel $\langle x^2 \rangle$.

32. The mapping from $Z_3[x]$ onto $Z_3[i]$ given by $\phi(f(x)) = f(i)$ is a ring homomorphism with Ker $\phi = \langle x^2 + 1 \rangle$. So, by Theorem 15.3, the ring $Z_3[x]/\langle x^2 + 1 \rangle$ is isomorphic to $Z_3[i]$.

33. First, observe that $\phi((0, 1)) = \phi((1, 1)) - \phi((1, 0)) = (1, 1) - (0, 1) = (1, 0)$. Then $\phi((a, b)) = a\phi((1, 0)) + b\phi((0, 1)) = a(0, 1) + b(1, 0) = (b, a)$.

34. Consider the mapping given by $(x, y) \to (x \bmod a, y \bmod b)$ and use Theorem 15.3.

35. Suppose that ϕ is a ring homomorphism from $Z \oplus Z$ to Z. Let $\phi((1, 0)) = a$ and $\phi((0, 1)) = b$. Since $\phi((x, y)) = \phi(x(1, 0) + y(0, 1)) = \phi(x(1, 0)) + \phi(y(1, 0)) = x\phi((1, 0)) + y\phi((0, 1)) = ax + by$ it suffices to determine a and b. Because $(1, 0)^2 = (1, 0)$ and $(0, 1)^2 = (0, 1)$, we know that $a^2 = a$ and $b^2 = b$. This means the $a = 0$ or 1 and $b = 0$ or 1. Thus there are four cases for (a, b):
(0,0) corresponds to $(x, y) \to 0$;
(1,0) corresponds to $(x, y) \to x$;
(0,1) corresponds to $(x, y) \to y$;
(1,1) corresponds to $(x, y) \to x + y$.
Each of the first is obviously a ring homomorphism. The last case is not because $2 = \phi((1, 1)) = \phi((1, 1)(1, 1)) = \phi((1, 1))\phi((1, 1)) = 4$.

36. $(n^2 + (n + 1)^2 + (n + 2)^2) \bmod 3 = 2$ while $k^2 = 2 \bmod 3$ has no solution.

37. Say $m = a_k a_{k-1} \cdots a_1 a_0$ and $n = b_k b_{k-1} \cdots b_1 b_0$. Then $m - n = (a_k - b_k)10^k + (a_{k-1} - b_{k-1})10^{k-1} + \cdots + (a_1 - b_1)10 + (a_0 - b_0)$. By the test for divisibility by 9 given in Example 8, $m - n$ is divisible by 9 provided that $a_k - b_k + a_{k-1} - b_{k-1} + \cdots + a_1 - b_1 + a_0 - b_0 = (a_k + a_{k-1} + \cdots + a_1 + a_0) - (b_k + b_{k-1} + \cdots + b_1 + b_0)$ is divisible by 9. But this difference is 0 since the second expression has the same terms as the first expression in some other order.

38. $(a_k a_{k-1} \dots a_1 a_0) \bmod 11 = (a_0 + 10a_1 + 10^2 a_2 + \cdots + 10^k a_k) \bmod 11 =$ $(a_0 - a_1 + a_2 - \cdots (-1)^k a_k) \bmod 11$.

39. Since the sum of the digits of the number is divisible by 9, so is the number (see Example 8); the test for divisibility by 11 given in Exercise 38 is not satisfied.

40. This follows directly from Exercise 74 of Chapter 10 and Exercise 11 of Chapter 0.

41. Let α be the homomorphism from Z to Z_3 given by $\alpha(n) = n \mod 3$. Then, noting that $\alpha(10^i) = \alpha(10)^i = 1^i = 1$, we have that $n = a_k a_{k-1} \cdots a_1 a_0 = a_k 10^k + a_{k-1} 10^{k-1} + \cdots + a_1 10 + a_0$ is divisible by 3 if and only if, modulo 3, $0 = \alpha(n) = \alpha(a_k) + \alpha(a_{k-1}) + \cdots + \alpha(a_1) + \alpha(a_0) = \alpha(a_k + a_{k-1} + \cdots + a_1 + a_0)$. But $\alpha(a_k + a_{k-1} + \cdots + a_1 + a_0) = 0 \mod 3$ is equivalent $a_k + a_{k-1} + \cdots + a_1 + a_0$ being divisible by 3.

42. $a_k a_{k-1} \cdots a_1 a_0 \bmod 4 = (a_k a_{k-1} \cdots a_1 a_0 - a_1 a_0) \bmod 4 + a_1 a_0 \bmod 4 = a_1 a_0 \bmod 4$.

43. Observe that the mapping ϕ from $Z_n[x]$ is isomorphic to Z_n given by, $\phi(f(x)) = f(0)$, is a ring-homomorphism onto Z_n with kernel $\langle x \rangle$ and use Theorem 15.3.

44. By Exercise 43 $Z[x]/\langle x \rangle \approx Z_n$. The statement now follows from Theorem 14.4 and the fact that Z_n is a field if and only if n is prime.

45. The ring homomorphism from $Z \oplus Z$ to Z given by $\phi(a, b) = a$ takes (1,0) to 1. Or define ϕ from Z_6 to Z_6 by $\phi(x) = 3x$ and let $R = Z_6$ and $S = \phi(Z_6)$. Then 3 is a zero-divisor in R and $\phi(3) = 3$ is the unity of S.

46. Since F^* is a group under multiplication any automorphism ϕ of F takes 1 to 1. By Corollary 3 of Theorem 15.3 we may assume the prime subfield is Z_p or Q. Then, by properties of isomorphisms, for any element n in Z_p or integer n in Q we have $\phi(n) = n$ and for any element m/n in Q we have $\phi(m/n) = \phi(m)/(\phi(n))^{-1} = mn^{-1} = m/n$.

47. Observe that 10 mod 3 = 1. So, $(2 \cdot 10^{75} + 2) \mod 3 = (2+2) \mod 3 = 1$ and $(10^{100} + 1) \mod 3 = (1+1) \mod 3 = 2 = -1 \mod 3$. Thus, $(2 \cdot 10^{75} + 2)^{100}$ $\mod 3 = 1^{100} \mod 3 = 1$ and $(10^{100} + 1)^{99} \mod 3 = 2^{99} \mod 3 = (-1)^{99}$ $\mod 3 = -1 \mod 3 = 2$.

48. By Exercise 27 we know that ϕ is the zero map or $\phi(m) = m$ for all integers m. Then, by property of Theorem 15.1, for all rationals m/n, we have $\phi(m/n) = \phi(mn^{-1}) = \phi(m)\phi(n^{-1}) = \phi(m)(\phi(n))^{-1} = mn^{-1} = m/n$.

49. By Theorem 13.3, the characteristic of R is the additive order of 1 and by property 6 of Theorem 15.1, the characteristic of S is the additive order of $\phi(1)$. Thus, by property 3 of Theorem 10.1, the characteristic of S divides the characteristic of R.

50. Use Exercise 49(a) of Chapter 13.

51. No. The kernel must be an ideal.

52. Use Exercise 37 of Chapter 14.

53. a. Suppose $ab \in \phi^{-1}(A)$. Then $\phi(ab) = \phi(a)\phi(b) \in A$, so that $a \in \phi^{-1}(A)$ or $b \in \phi^{-1}(A)$.

b. Let Φ be the homomorphism from R to S/A given by $\Phi(r) = \phi(r) + A$. Then $\phi^{-1}(A) = \text{Ker}\ \Phi$ and, by Theorem 15.3, $R/\text{Ker}\ \Phi \approx S/A$. So, $\phi^{-1}(A)$ is maximal.

54. If $\phi^{-1}(A) = \langle a \rangle$, then $A = \langle \phi(a) \rangle$.

55. **a.** Since $\phi((a,b) + (a',b')) = \phi((a+a', b+b')) = a + a' = \phi((a,b)) + \phi((a',b'))$, ϕ preserves addition. Also,
$\phi((a,b)(a',b')) = \phi((aa',bb')) = aa' = \phi((a,b))\phi((a',b'))$ so ϕ preserves multiplication.

b. $\phi(a) = \phi(b)$ implies that $(a,0) = (b,0)$, which implies that $a = b$.
$\phi(a+b) = (a+b,0) = (a,0) + (b,0) = \phi(a) + \phi(b)$. Also,
$\phi(ab) = (ab,0) = (a,0)(b,0) = \phi(a)\phi(b)$.

c. Define ϕ by $\phi(r,s) = (s,r)$. By Exercise 7 in Chapter 8, ϕ is one-to-one and preserves addition. Since
$\phi((r,s)(r',s')) = \phi((rr',ss')) = (ss',rr') = (s,r)(s',r') = \phi((r,s))\phi((r',s'))$
multiplication is also preserved.

56. Note that $I \oplus S$ is the pull back of the homomorphism from $R \oplus S$ onto R given by $(a,b) \to a$ (see Exercises 55 and 53).

57. The mapping $\phi(x) = (x \bmod m, x \bmod n)$ from Z_{mn} to $Z_m \oplus Z_n$ is a ring isomorphism.

58. Suppose that ϕ is a ring isomorphism from mZ onto nZ. Observe that by property 1 of Theorem 15.1 we have $\phi(mm) = m\phi(m)$ and $\phi(mm) = \phi(m)^2$. This gives us that $\phi(m) = m$.

59. First, note that $\phi(1) = 1$ implies that $\phi(m) = \phi(m \cdot 1) = m\phi(1) = m$. Now let $\phi(\sqrt[3]{2}) = a$. Then $2 = \phi(2) = \phi(\sqrt[3]{2}^3) = (\phi(\sqrt[3]{2}))^3$ and therefore $\phi(\sqrt[3]{2}) = \sqrt[3]{2}$.

60. First show that any automorphism ϕ of $\mathbf{R}$ acts as the identity map on the rationals (compare with Exercise 68 of Chapter 68) Then show that if $a < b$ then $\phi(a) < \phi(b)$ (see Exercise 71 of Chapter 6). Next suppose that there is some a such that $\phi(a) \neq a$. Say, $a < \phi(a)$. Pick a rational number r such that $a < r < \phi(a)$. Then $\phi(a) < \phi(r) = r$, a contradiction. A similar argument applies if $\phi(a) < a$.

61. By Exercise 52, every non-trivial ring homomorphism from $\mathbf{R}$ to $\mathbf{R}$ is an automorphism of $\mathbf{R}$. And by Exercise 60 the only automorphism of $\mathbf{R}$ is the identity.

62. To check that multiplication is operation preserving, observe that $xy \to a(xy) = a^2xy = axay$. For the second part take $m = 4, n = 6$ and $a = 4$. Then $0 = \phi(0) = \phi(2 \cdot 2)$ but $\phi(2)\phi(2) = 2 \cdot 2 = 4$.

63. If $a/b = a'/b'$ and $c/d = c'/d'$, then $ab' = ba'$ and $cd' = dc'$. So, $acb'd' = (ab')(cd') = (ba')(dc') = bda'c'$. Thus, $ac/bd = a'c'/b'd'$ and therefore $(a/b)(c/d) = (a'/b')(c'/d')$.

64. Suppose that $\sqrt{2}$ maps to $a + b\sqrt{5}$. Then 2 maps to $(a+b\sqrt{5})^2 = a^2 + 2ab\sqrt{5} + b^2$. Solving for $\sqrt{5}$ we get a rational number, which is a contradiction.
Alternate proof. First observe that $1 \to 1$ so that $2 \to 2$. Suppose $\sqrt{2} \to x$. Then $2 = \sqrt{2}\sqrt{2} \to x^2$. This implies that $x = \pm\sqrt{2}$, which are not in $Q(\sqrt{5})$.

65. Let F be the field of quotients of $Z[i]$. By definition $F = \{(a+bi)/(c+di) \mid a,b,c,d \in Z\}$. Since F is a field that contains Z and i, we know that $Q[i] \subseteq F$. But for any $(a+bi)/(c+di)$ in F we have $\frac{a+bi}{c+di} = \frac{a+bi}{c+di}\frac{c-di}{c-di} = \frac{(ac+bd)+(bc-ad)i}{c^2+d^2} = \frac{ac+bd}{c^2+d^2} + \frac{(bc-ad)i}{c^2+d^2} \in Q[i]$.

66. Map $[a/b]$ to ab^{-1}.

67. The subfield of E is $\{ab^{-1} \mid a,b \in D, b \neq 0\}$. Define ϕ by $\phi(ab^{-1}) = a/b$. Then $\phi(ab^{-1} + cd^{-1}) = \phi((ad+bc)(bd)^{-1})) = (ad+bc)/bd = ad/bd + bc/bd = a/b + c/d = \phi(ab^{-1}) + \phi(cd^{-1})$. Also, $\phi((ab^{-1})(cd^{-1})) = \phi(acb^{-1}d^{-1}) = \phi((ac)(bd)^{-1}) = ac/bd = (a/b)(c/d) = \phi(ab^{-1})\phi(cd^{-1})$.

68. Zero-divisors do not have multiplicative inverses.

69. Reflexive and symmetric properties follow from the commutativity of D. For transitivity, assume $a/b \equiv c/d$ and $c/d \equiv e/f$. Then $adf = (bc)f = b(cf) = bde$, and cancellation yields $af = be$.

70. The set of even integers is a subring of the rationals.

71. Let ϕ be the mapping from T to Q given by $\phi(ab^{-1}) = a/b$. Now see Exercise 67.

72. From Example 2 we have that the mapping that sends $x + yi + \langle a+bi \rangle$ to $x + yi + \langle a-bi \rangle$ is an isomorphism.

73. Let $a_n x^n + a_{n-1}x^{n-1} + \cdots + a_0 \in \mathbf{R}[x]$ and suppose that $f(a+bi) = 0$. Then $a_n(a+bi)^n + a_{n-1}(a+bi)^{n-1} + \cdots + a_0 = 0$. By Example 2, the mapping ϕ from $\mathbf{C}$ to itself given by $\phi(a+bi) = a-bi$ is a ring isomorphism. So, by property 1 of Theorem 10.1,
$0 = \phi(0) = \phi(a_n(a+bi)^n + a_{n-1}(a+bi)^{n-1} + \cdots + a_0) = \phi(a_n)\phi((a+bi))^n + \phi(a_{n-1})\phi((a+bi))^{n-1} + \cdots + \phi(a_0) = a_n(a-bi)^n + a_{n-1}(a-bi)^{n-1} + \cdots + a_0 = f(a-bi)$.

74. **a.** Apply the definition.

b. Ker $\phi = \left\{ \begin{bmatrix} a & b \\ c & d \end{bmatrix} \middle| \; a \in Z \right\}$.

c. Use Theorem 15.3.

d. Yes, by Theorem 14.3.

e. No, by Theorem 14.4.

75. Certainly, the unity 1 is contained in every subfield. So, if a field has characteristic p, the subfield $\{0, 1, \ldots, p-1\}$ is contained in every subfield. If a field has characteristic 0, then $\{(m \cdot 1)(n \cdot 1)^{-1} \mid m, n \in Z, n \neq 0\}$ is a subfield contained in every subfield. This subfield is isomorphic to Q [map $(m \cdot 1)(n \cdot 1)^{-1}$ to m/n].

76. By part 5 of Theorem 6.1, the only possible isomorphism is given by $1 \to n$. If this mapping is an isomorphism, then $1 = 1^2 \to n^2$. So $n^2 = n \bmod 2n$ and it follows that n is odd. Now suppose n is odd. Then $n(n-1)$ is divisible by $2n$ and $n^2 = n \bmod 2n$. This guarantees that $1 \to n$ is an isomorphism.

77. $x \to 6x$ is the only possibility.

78. By Exercise 5 it suffices to find an element of order 3 in Z_6 that is an idempotent. The only choice is 4.

79. Say ϕ is a ring homomorphism from $2Z$ to $2Z$ and $\phi(2) = a$. Then, $\phi(4) = \phi(2) + \phi(2) = 2a$ and $\phi(4) = \phi(2 \cdot 2) = \phi(2)^2 = a^2$. So, $2a = a^2$, which implies that $a = 0$ or $a = 2$.
For the general case, let $\phi(n) = a$. Then $\phi(n \cdot n) = n\phi(n) = na$ and $\phi(n \cdot n) = \phi(n)^2 = a^2$. So, $a = n^2a$, which implies that $a = 0$ or $n^2 = 1$, which implies that $n = 1$. In words, the only ring homomorphisms from nZ to nZ are the identity and the zero map.

CHAPTER 16

Polynomial Rings

1. $$\begin{aligned} f+g &= 3x^4+2x^3+2x+2 \\ f\cdot g &= 2x^7+3x^6+x^5+2x^4+3x^2+2x+2. \end{aligned}$$

2. Let $f(x)=x^4+x$ and $g(x)=x^2+x$. Then $f(0)=0=g(0); f(1)=2=g(1); f(2)=0=g(2)$.

3. The zeros are 1, 2, 4, 5.

4. Since R is isomorphic to the subring of constant polynomials, char$R \leq$ char $R[x]$. On the other hand, char $R=c$ implies $c(a_nx^n+\cdots+a_0)=(ca_n)x^n+\cdots+(ca_0)=0$.

5. The only place in the proof of Theorem 16.2 and its corollaries that uses the fact the coefficients were from a field is where we used the multiplicative inverse of lead coefficient b_m of $g(x)$.

6. x^2, x^2+1, x^2+x, x^2+x+1. No two define the same function from Z_2 to Z_2.

7. Note that the functions defined by $f(x)=x^3, x^5, x^7, \ldots$, are the same as the one defined by $f(x)=x$ and the ones defined by $f(x)=x^4, x^6, x^8, \ldots$, are the same as the one defined by $f(x)=x^2$. So all such terms may be replaced by x and x^2. In the general case note that by Fermat's Little Theorem (Corollary 5 to Theorem 7.1) the function from Z_p to Z_p defined by $g(x)=x^p$ is the same as the function $f(x)=x$ from Z_p to Z_p. So, every polynomial function with coefficients from Z_p can be written in the form $a_{p-1}x^{p-1}+\cdots+a_0$ where $a_{p-1},\ldots,a_0 \in Z_p$.

8. In the ring functions from Z_3 to Z_3 the functions $f(x)=x^3$ and $g(x)=x$ are equal. So, $0=f(x)-g(x)=x^3-x=x(x^2-x)$ but $h(x)=x$ and $k(x)=x^2-x$ are not zero functions. This does not contradict Theorem 16.1 because the ring of functions from Z_3 to Z_3 is not the same as the ring of polynomials from Z_3 to Z_3.

9. $(x-1)^2(x-2)$.

10. There are 2^n polynomials over Z_2. There are 4 polynomial functions from Z_2 to Z_2.

11. $4x^2+3x+6$ is the quotient and $6x+2$ is the remainder.

12. $(x-i)(x+i)(x-(2+i))(x-(2-i))$

13. Let $f(x)$, $g(x) \in R[x]$. By inserting terms with the coefficient 0 we may write

$$f(x)=a_nx^n+\cdots+a_0 \quad \text{and} \quad g(x)=b_nx^n+\cdots+b_0.$$

Then

$$\begin{aligned} \overline{\phi}(f(x)+g(x)) &= \phi(a_n+b_n)x^n+\cdots+\phi(a_0+b_0) \\ &= (\phi(a_n)+\phi(b_n))x^n+\cdots+\phi(a_0)+\phi(b_0) \\ &= (\phi(a_n)x^n+\cdots+\phi(a_0))+(\phi(b_n)x^n+\cdots+\phi(b_0)) \\ &= \overline{\phi}(f(x))+\overline{\phi}(g(x)). \end{aligned}$$

Multiplication is done similarly.

14. Use Exercise 13 and observe that $\phi(a_n)x^n + \cdots + \phi(a_0) = 0$ if and only if $\phi(a_n) = 0, \ldots, \phi(a_0) = 0$. Since ϕ is an isomorphism, this holds if and only if $a_n = 0, \ldots, a_0 = 0$.

15. Note that $(2x^n + 1)^2 = 1$ and $(2x^n)^2 = 0$ for all n.

16. If $f(x) \neq g(x)$, then $f(x) - g(x)$ would be a polynomial with degree at most 3 with 4 zeros, contradicting Theorem 16.3. In general, if $f(x)$ and $g(x)$ are polynomials over a field of degree at most n and there are more than n values of a such that $f(a) = g(a)$, then $f(x) = g(x)$.

17. Observe that $(2x+1)(2x+1) = 4x^2 + 4x + 1 = 1$. So, $2x + 1$ is its own inverse.

18. Let a be a zero-divisor. Then $f(x) = ax$ has degree 1 with zeros 0 and a.

19. If $f(x) = a_n x^n + \cdots + a_0$ and $g(x) = b_m x^m + \cdots + b_0$, then $f(x) \cdot g(x) = a_n b_m x^{m+n} + \cdots + a_0 b_0$ and $a_n b_m \neq 0$ when $a_n \neq 0$ and $b_m \neq 0$.

20. Suppose that $f(x) = a_n x^n + \cdots + a_0$ from $Z_4[x]$ is an idempotent where $a_n \neq 0$ and $n \geq 1$. Because the lead term of $f(x)^2$ is $a_n^2 x^{2n}$ and the lead term of $f(x)$ is $a_n x^n$ we have $a_n^2 = 0$. Thus $a_n = 2$. Comparing the coefficients of x^n in both $f(x)^2$ and $f(x)$ we obtain $2a_n a_0 = a_n$ and therefore $a_n = 0$. This is a contradiction.

21. Let m be the multiplicity of b in $q(x)$. Then we may write $f(x) = (x-a)^n (x-b)^m q'(x)$ where $q'(x)$ is in $F[x]$ and $q'(b) \neq 0$. This means that b is a zero of $f(x)$ of multiplicity at least m. If b is a zero of $f(x)$ of multiplicity greater than m, then b is a zero of $g(x) = f(x)/(x-b)^m = (x-a)^n q'(x)$. But then $0 = g(b) = (b-a)^n q'(b)$ and therefore $q'(b) = 0$, which is a contradiction.

22. 1 is a zero when the field has characteristic 2. Over all other fields 2 is a zero.

23. Let $f(x), g(x) \in R[x]$. By adding coefficients with coefficient 0 in the front we can write $f(x) = a_n x^n + a_{n-1}x^{n-1} + \cdots + a_1 x + a_0$ and $g(x) = b_n x^n + b_{n-1}x^{n-1} + \cdots + b_1 x + b_0$. Then $\phi(f(x)+g(x)) = \phi(a_n x^n + a_{n-1}x^{n-1} + \cdots + a_1 x + a_0 + b_n x^n + b_{n-1}x^{n-1} + \cdots + b_1 x + b_0) = \phi((a_n + b_n)x^n + (a_{n-1} + b_{n-1}x^{n-1} + \cdots + (a_1 + b_1)x + a_0 + b_0) = (a_n + b_n)r^n + (a_{n-1} + b_{n-1})r^{n-1} + \cdots + (a_1 + b_1)r + a_0 + b_0) = a_n r^n + a_{n-1}r^{n-1} + \cdots + a_1 r + a_0 + b_n r^n + b_{n-1}r^{n-1} + \cdots + b_1 r + b_0 = f(r) + g(r) = \phi(f(x)) + \phi(g(x))$. The analogous argument works for multiplication.

24. By Example 2 of Chapter 15 $f(i) = 0$ implies $f(-i) = 0$. Now observe that $f(i^3) = f(i^2 i) = f(-i)$.

25. Since $f(2) = 16 - 4 - 2 = 10$, $p = 2$ or 5.

26. Multiple $a_n b^n + a_{n-1}b^{n-1} + \cdots + a_1 b + a_0 = 0$ by b^{-n}.

27. By observation, $U(2)$ and $U(3)$ are cyclic. If $U(p)$ is not cyclic and $p > 3$, then by the Fundamental Theorem of Finite Abelian groups there is some prime q such that $U(p)$ has a subgroup isomorphic to $Z_q \oplus Z_q$. But the polynomial $x^q - 1$ in $Z_q[x]$ has $q^2 - 1$ zeros, which contradicts Theorem 16.3.

28. For every positive integer m the element $2x^{2m} - 1$ has multiplicative order 2.

29. In Z_{10}, let $f(x) = 5x$. Then $0, 2, 4, 6, 8$ are zeros.

30. Let $n = ab$ where $a > 1$, $b > 1$). If $a = b$ denote the remaining $n-2$ non-zero elements of Z_n by $a_3, \ldots, a_n$. Then every element of Z_n is a zero of $f(x) = ax(x-a_3)\cdots(x-a_n)$ and $f(x)$ has degree $n-1$. If $a \neq b$, denote the remaining $n-3$ non-zero elements of Z_n by $a_4, \ldots, a_n$. Then every element of Z_n is a zero of $f(x) = ax(x-a)(x-a_4)\cdots(x-a_n)$ and $f(x)$ has degree $n-1$.

31. If $(f(x)/g(x))^2 = x$, then $x^2(k(x))^2 = x(g(x))^2$. But the right side has even degree whereas the left side has odd degree.
Alternate solution. Say $(f(x)/g(x))^2 = x$. We may assume that $f(x)$ and $g(x)$ have no common factor for, if so, we can cancel them. Since $(f(x))^2 = x(g(x))^2$ we see that $f(0) = 0$. Thus, $f(x)$ has the form $xk(x)$. Then $x^2(k(x))^2 = x(g(x))^2$ and therefore $x(k(x))^2 = (g(x))^2$. This implies that $g(0) = 0$. But then $f(x)$ and $g(x)$ have x as a common factor.

32. $f(x) = x(x-1)(x-2)(x-3)(x-4)+1$.

33. Suppose that $f(x) \in D[x]$ has degree at least 1 and there is a $g(x) \in D[x]$ such that $f(x)g(x) = 1$. Then by Exercise 19 $0 = \deg f(x)g(x) = \deg\ (f(x) + \deg g(x) \geq 1$.

34. $(x-1)^2(x+1)$.

35. $2x(x-1)$ has zeros 0, 1, 2, and 4.

36. This follows directly from Theorem 16.5 and Exercise 19.

37. By Theorem 16.5, $g(x) = (x-1)(x-2)$.

38. Suppose that $f(x) = a_m x^m + \cdots + a_0$ is an idempotent where $a_m \neq 0$ and $m \geq 1$. Because the lead term of $(f(x))^2$ is $a_m^2 x^{2m}$ and the lead term of $f(x)$ is $a_m x^m$ we have $a_m^2 = 0$ and therefore $a_m = 0$.

39. First, note that $-1 = 16$ is zero. Since $x^9 + 1 = 0$ implies $x^{18} = 1$ in the group $U(17)$, for any solution a of $x^9 + 1 = 0$ in the group $U(17)$, we know that $|a|$ must divide 18 and $|a|$ must divide $|U(17)| = 16$. This gives us $|a| = 2$ and $a^9 + 1 = a + 1$ so that $a = -1$. Because $U(17)$ is cyclic, 16 is the unique element of order 2.

40. The ring homomorphism from ϕ from $Q[x]$ to $Q(i) = \{a + bi \mid a, b \in Q\}$ given by $\phi(f(x)) = f(i)$ has kernel $\langle x^2+1\rangle$. So, by Theorem 15.3 $Q[x]/\langle x^2+1\rangle$ is isomorphic to $Q(i)$.

41. Since -1 is a zero of $x^{25}+1$, $x+1$ is a factor. Suppose that $x^{25}+1 = (x+1)^2 g(x)$ for some $g(x) \in Z_{37}[x]$. Then the derivative $f'(x) = 25x^{24} = (x+1)^2 g'(x) + g(x)2(x+1)$. This gives $f'(-1) = 25 = 0$, which is false.

42. Let $f(x)$ be a non-constant polynomial of minimum degree with the stated property. Then $g(x) = f(x) - 2$ has five zeros and since Z_5 is a field, $g(x)$ has degree 5 and has the same degree as $f(x)$.

43. Since $F[x]$ is a PID, $\langle f(x), g(x)\rangle = \langle a(x)\rangle$ for some $a(x) \in F[x]$. Thus $a(x)$ divides both $f(x)$ and $g(x)$. This means that $a(x)$ is a constant. So, by Exercise 17 in Chapter 14, $\langle f(x), g(x)\rangle = F[x]$. Thus, $1 \in \langle f(x), g(x)\rangle$.

44. Suppose that $d_1(x)$ and $d_2(x)$ are both greatest common divisors of $f(x)$ and $g(x)$ and monic. Then, By Theorem 16.2, $\langle d_1(x)\rangle = \langle d_2(x)\rangle$. Thus, $d_1(x) = cd_2(x)$ for some c in F. But then, $c = 1$.

45. Suppose that $I = \langle f(x) \rangle$. Then there is some $g(x) \in Z[x]$ such that $2 = f(x)g(x)$. This implies that $f(x) = \pm 2$. But $x + 2 \in I$ and is not in $\langle 2 \rangle$.

46. If $f(x) = g(x)$ for infinitely many elements of F then $h(x) = f(x) - g(x)$ has infinitely many zeros. So, by Exercise 16, $h(x) = 0$.

47. If $f(x) \neq g(x)$, then $\deg[f(x) - g(x)] < \deg p(x)$. But the minimum degree of any member of $\langle p(x) \rangle$ is $\deg p(x)$. So, $f(x) - g(x)$ does not have a degree. This means that $f(x) - g(x) = 0$.

48. We start with $(x - 1/2)(x + 1/3)$ and clear fractions to obtain $(6x - 3)(6x + 2)$ as one possible solution.

49. For any positive integer k that are at most k zeros of $x^k - 1$. So, there are at most k elements in the field that are solutions to $x^k = 1$.

50. We know $f(x)$ can be written in the form $(x - a)g(x)$. Then $f'(x) = (x - a)g'(x) + g(x)$ and $g(a) = f'(a) \neq 0$. So, $x - a$ is not a factor of $g(x)$.

51. The proof given for Theorem 16.2 with $g(x) = x - a$ is valid over any commutative ring with unity. Moreover, the proofs for Corollaries 1 and 2 of Theorem 16.2 are also valid over any commutative ring with unity.

52. Notice that the proof of the division algorithm holds for integral domains when $g(x)$ has the form $x - a$. Likewise the proofs of the Factor Theorem and Corollary 3 of Theorem 16.2 hold.

53. Observe that $f(x) \in I$ if and only if $f(1) = 0$. Then if f and g belong to I and h belongs to $F[x]$, we have $(f - g)(1) = f(1) - g(1) = 0 - 0$ and $(hf)(1) = h(1)f(1) = h(1) \cdot 0 = 0$. So, I is an ideal. By Theorem 16.5, $I = \langle x - 1 \rangle$.

54. Use the Factor Theorem.

55. For each positive integer k observe that $(px^k + 1)(-px^k + 1) = 1$.

56. Every element in the ideal $\langle x^3 - x \rangle$ satisfies the condition.

57. For any a in $U(p)$, $a^{p-1} = 1$, so every member of $U(p)$ is a zero of $x^{p-1} - 1$. From the Factor Theorem (Corollary 2 of Theorem 16.2) we obtain that $g(x) = (x - 1)(x - 2) \cdots (x - (p - 1))$ is a factor of $x^{p-1} - 1$. Since both $g(x)$ and $x^{p-1} - 1$ have lead coefficient 1, the same degree, and their difference has $p - 1$ zeros, their difference must be 0 (for otherwise their difference would be a polynomial of degree less than $p - 1$ that had $p - 1$ zeros).

58. When n is prime, use Exercise 57. The case $n = 4$ is done by observation. When n is composite and greater than 4, $(n - 1)! \bmod n = 0$.

59. By Exercise 58, $(p - 1)! \bmod p = p - 1$. Since $p - 1 = -1$, mod p we have $-(p - 2)! = -1$ and the statement follows.

60. The problem is to solve $98! = x \bmod 101$ for x. By Exercise 59, modulo 101, we have $1 = 99! = (-2)98! = -2x$. Then, by observation, $x = -51 = 50$.

61. Let $x^{48} + x^{21} + a$. Since $x + 4 = x - 1 \in Z_5[x]$ from the Factor Theorem we need only find an a in Z_5 such that $f(1) = 1 + 1 + a = 0$. So $a = 3$.

62. First, note that $x - 1 \in \text{Ker } \phi$. Let $f(x) \in \text{Ker } \phi$. Then by Theorem 16.5, $x - 1$ is a factor of $f(x)$. So $\text{Ker } \phi = \langle x - 1 \rangle$. By Theorem 15.3, $Q[x]/\text{Ker } \phi$ is isomorphic Q.

63. $\mathbf{C}(x)$ (field of quotients of $\mathbf{C}[x]$).

64. This follows directly from the definitions.

65. Note that $I = \langle 2 \rangle$ is maximal in Z but $I[x]$ is not maximal in $Z[x]$ since $I[x]$ is properly contained in the ideal $\{f(x) \in Z[x] \mid f(0) \text{ is even}\}$.

66. That $I[x]$ is an ideal is straightforward. To prove that $I[x]$ is prime let $f(x) = a_m x^m + \cdots + a_0$ and $g(x) = b_n x^n + \cdots + b_0$ and suppose $f(x)g(x) \in I[x]$. By filling in with coefficients of 0 we may assume that $m = n$. We must show that all $a_i \in I$ or all $b_i \in I$. Suppose some $b_i \notin I$ and let k be the least integer such that $b_k \notin I$. The coefficient x^k in $f(x)g(x)$ is $a_k b_0 + a_{k-1}b_1 + \cdots + a_0 b_k$ and belongs to I. Thus $a_0 b_k \in I$ and therefore $a_0 \in I$. The coefficient of x^{k+1} in $f(x)g(x)$ is $a_{k+1}b_0 + a_k b_1 + \cdots + a_1 b_k + a_0 b_{k+1} \in I$. Thus, $a_1 b_k \in I$ and therefore $a_1 \in I$. Continuing in this fashion we obtain all $a_i \in I$.

67. A solution to $x^{25} - 1 = 0$ in Z_{37} is a solution to $x^{25} = 1$ in $U(37)$. So, by Corollary 2 of Theorem 4.1, $|x|$ divides 25. Moreover, we must also have that $|x|$ divides $|U(37)| = 36$. So, $|x| = 1$ and therefore $x = 1$.

68. Mimic Example 3.

69. By the Factor Theorem (Corollary 2 of Theorem 16.2) we may write $f(x) = (x - a)g(x)$. Then $f'(x) = (x - a)g'(x) + g(x)$. Thus, $g(a) = 0$ and by the Factor Theorem, $x - a$ is a factor of $g(x)$.

70. Observe that the zero polynomial is in I. If f and $g \in I$, then $(f - g)(a) = f(a) - g(a) = 0$ for all a. If $f \in I$ and $g \in F[x]$, then $(gf)(a) = g(a)f(a) = 0$ for all a.
If $F = \{a_1, a_2, \ldots, a_n\}$ then for every non-negative integer k the polynomial $x^k(x - a_1)(x - a_2)\cdots(x - a_n)$ is in I. If F is infinite and $f(x) \in I$, then $f(x)$ has infinitely many zeros. So, by Theorem 16.3, $f(x) = 0$.

71. Say $\deg g(x) = m, \deg h(x) = n$, and $g(x)$ has leading coefficient a. Let $k(x) = g(x) - ax^{m-n}h(x)$. Then $\deg k(x) < \deg g(x)$ and $h(x)$ divides $k(x)$ in $Z[x]$ by induction. So, $h(x)$ divides $k(x) + ax^{m-n}h(x) = g(x)$ in $Z[x]$.

72. The mapping $\phi(f(x)) = f(x^2)$ is a ring-isomorphism from $R[x]$ onto $R[x^2]$.

73. If $f(x)$ takes on only finitely many values, then there is at least one a in Z with the property that $f(x) = a$ for infinitely many x in Z. But then $g(x) = f(x) - a$ has infinitely many zeros. This contradicts Corollary 3 of Theorem 16.2.

74. By Theorem 16.5, $I = \langle x(x - 1) \rangle$. In general, if $A = \{a_1, a_2, \ldots, a_n\}$ is any finite subset of a field F and $I = \{f(x) \in F[x] \mid f(a_i) = 0 \text{ for all } a_i \in A\}$, then $I = \langle (x - a_1)(x - a_2)\cdots(x - a_n) \rangle$.

75. Let ϕ be a ring homomorphism from Z onto a field and let Ker $\phi = nZ$. Then by Theorem 15.3 we have $Z/nZ \approx Z_n$ is a field. From Theorem 14.3 we have that nZ is a prime ideal of Z, and from Example 14 in Chapter 14, we know that n is a prime.

76. Say $f(x) = a_n x^n + a_{n-1}x^{n-1} + \cdots + a_1 x + a_0$. Then using the fact that $a_i^p = a_i$ for all i and Exercise 49 of Chapter 13, we have
$f(b^p) = a_n^p(b^p)^n + a_{n-1}^p(b^p)^{n-1} + \cdots + a_1^p b^p + a_0^p =$
$a_n^p(b^n)^p + a_{n-1}^p(b^{n-1})^p + \cdots + a_1^p b + a_0^p = (a_n b^n + a_{n-1}b^{n-1} + \cdots + a_1 b + a_0)^p = f(b)^p = 0.$

77. Let $f(x) = a_n x^n + a_{n-1}x^{n-1} + \cdots + a_1 x + a_0$ where $a_0, a_1, \ldots a_n$ are odd integers and assume that p/q is a zero of $f(x)$ where p and q are integers and n is even. We may assume that p and q are not both even. Substituting p/q for x and clearing fractions we have $a_n p^n + a_{n-1}p^{n-1}q + \cdots + a_1 pq^{n-1} = -a_0 q^n$. If both p and q are odd, then the left side is even since it has an even number of odd terms. If p is even and q is odd, then the left side is even and the left side odd. If p is odd and q is even, then the first term on the left is odd and all the other terms on the left are even. So, the left side is odd and the right side is even. Thus, in each case we have a contradiction.

78. Since $x + 4 = x - 3$ in $Z_7[x]$, we have by the Remainder Theorem that the remainder is 3^{51} mod 7. Since 3 is in $U(7)$ we also know that $3^6 = 1$ mod 7. Thus, 3^{51} mod 7 = $3^{48}3^3$ mod 7 = 6.

79. By the Division Algorithm (Theorem 16.2) we may write $x^{43} = (x^2 + x + 1)q(x) + r(x)$ where $r(x) = 0$ or deg $r(x) < 2$. Thus, $r(x)$ has the form $cx + d$. Then $x^{43} - cx - d$ is divisible by $x^2 + x + 1$. Finally, let $a = -c$ and $b = -d$.

80. $2x(x - 4)$

CHAPTER 17

Factorization of Polynomials

1. By Theorem 17.1, $f(x)$ is irreducible over **R**. Over **C** we have $2x^2+4=2(x^2+2)=2(x+\sqrt{2}i)(x-\sqrt{2}i)$.

2. $f(x)$ factors over D as $ah(x)$ where a is not a unit.

3. If $f(x)$ is not primitive, then $f(x)=ag(x)$, where a is an integer greater than 1. Then a is not a unit in $Z[x]$ and $f(x)$ is reducible.

4. Say $r=p/q$ where p and q are relatively prime. Viewing $f(x)$ as an element of $Q[x]$ we have from the Factor Theorem (Corollary 2 of Theorem 16.2) that $f(p/q)=0$. Clearing fractions and collecting all terms and isolating the p^n term on one side we see that q divides p^n. Using the fact that p and q are relatively prime, we conclude that $q=1$.

5. **a.** If $f(x)=g(x)h(x)$, then $af(x)=ag(x)h(x)$.
   b. If $f(x)=g(x)h(x)$, then $f(ax)=g(ax)h(ax)$.
   c. If $f(x)=g(x)h(x)$, then $f(x+a)=g(x+a)h(x+a)$.
   d. Let $f(x)=8x^3-6x+1$. Then $f(x+1)=8(x+1)^3-6(x+1)+1=$ $8x^3+24x^2+24x+8-6x-6+1=8x^3+24x^2-18x+3$.
   By Eisenstein's Criterion (Theorem 17.4), $f(x+1)$ is irreducible over Q and by part c, $f(x)$ is irreducible over Q.

6. Use Exercise 5(a).

7. Suppose that $r+1/r=2k+1$ where k is an integer. Then $r^2-2kr-r+1=0$. It follows from Exercise 4 of this chapter that r is an integer. But the mod 2 irreducibility test shows that the polynomial $x^2-(2k+1)x+1$ is irreducible over Q and an irreducible quadratic polynomial cannot have a zero in Q.

8. Use the Mod p Test with 3, 2, 5, and 7 respectively.

9. Use Exercise 5a and clear fractions.

10. By Corollary 1 of Theorem 17.5 we know the set is a field. To see that it has p^n elements, note that any element

$$g(x)+\langle f(x)\rangle = f(x)q(x)+r(x)+\langle f(x)\rangle = r(x)+\langle f(x)\rangle$$

where $r(x)=0$ or deg $r(x)<n$. So, all cosets have the form

$$a_{n-1}x^{n-1}+\cdots+a_0+\langle f(x)\rangle$$

and they are all distinct.

11. It follows from Theorem 17.1 that $p(x)=x^2+x+1$ is irreducible over Z_5. Then, from Corollary 1 of Theorem 17.5, we know that $Z_5[x]/\langle p(x)\rangle$ is a field. To see that this field has order 25, note that if $f(x)+\langle p(x)\rangle$ is any element of $Z_5[x]/\langle p(x)\rangle$, then by the Division Algorithm (Theorem 16.2) we may write $f(x)+\langle p(x)\rangle$ in the form

$p(x)q(x) + ax + b + \langle p(x) \rangle = ax + b + \langle p(x) \rangle$. Moreover, $ax + b + \langle p(x) \rangle = cx + d + \langle p(x) \rangle$ only if $a = c$ and $b = d$, since $(a - c)x + b - d$ is divisible by $\langle p(x) \rangle$ only when it is 0. So, $Z_5[x]/\langle p(x) \rangle$ has order 25.

12. Find an irreducible cubic over Z_3 and mimic Example 10. One such cubic is $x^3 + x^2 + 2$ (by the Mod 3 Test).

13. **a.** Irreducible by Eisenstein

    b. Irreducible by the Mod 2 Test (but be sure to check for quadratic factors as well as linear)

    c. Irreducible by Eisenstein

    d. Irreducible by the Mod 2 Test

    e. Irreducible by Eisenstein (after clearing fractions)

14. Note that -1 is a zero. No, since 4 is not a prime.

15. x; $x + 1$

16. Note that $|F^*| = 7$ and $|a|$ divides 7. Yes, for $|F| = 32$, but no for $|F| = 16$. (Because 31 is prime but 15 is not.)

17. $f(x)$ is irreducible over Q. Nothing.

18. If $f(x)$ is reducible over Z_2 and does not have 0 or 1 as a zero then it must factor as an irreducible quadratic and an irreducible cubic. But the only irreducible quadratic over Z_2 is $x^2 + x + 1$.

19. We first observe that the multiplicative group of the field has order 48. So, the nonzero elements have orders that divide 48. To find $|x|$ we use the fact that mod I, $x^2 = -2 = 5 \bmod 7$. Since $x^2 = 5$, we have that $x^3 = 5x$ and $x^6 = 4x^2 = -8 = -1$. So, $x^{12} = 1$. To find $|x + 1|$, we note that $(x + 1)^2 = x^2 + 2x + 1 = 2x - 1$; $(x + 1)^4 = (2x - 1)^2 = 4x^2 - 4x + 1 = -4x$; $(x + 1)^8 = 2x^2 = 3$. Since observe that $|3| = 6$. So, $|x + 1| = 48$. Because $|x| = 12$, we know that $x^{-1} = x^{11} = x^6 x^3 x^3 = -1(5x)(5x) = -4x = 3x$. To check this we note that $x(3x) = 3x^2 = -6 = 1$.

20. Let $f(x) = g(x)h(x)$ where deg $g(x) <$ deg $f(x)$ and deg $h(x) <$ deg $f(x)$.

21. Let $f(x) = x^4 + 1$ and $g(x) = f(x + 1) = x^4 + 4x^3 + 6x^2 + 4x + 2$. Then $f(x)$ is irreducible over Q if $g(x)$ is. Eisenstein's Criterion shows that $g(x)$ is irreducible over Q.
    Alternate proof. Since $x^4 + 1$ has no real zeros, the only possible factorizations over Z are $x^4 + 1 = (x^2 + ax + 1)(x^2 + bx + 1)$ or $x^4 + 1 = (x^2 + ax - 1)(x^2 + bx - 1)$. Evaluating $x^4 + 1 = (x^2 + ax + 1)(x^2 + bx + 1)$ at 1 gives us $2 = (a + 2)(b + 2)$. So, one of a or b is 0. But long division shows $x^2 + 1$ is not a factor of $x^4 + 1$. Evaluating $x^4 + 1 = (x^2 + ax - 1)(x^2 + bx - 1)$ at 1 gives us $2 = ab$. So, one of a or b is 1. But long division shows $x^2 + x - 1$ is not a factor of $x^4 + 1$.

22. Observe that if $f(x) = p(x)q(x)$, then $f(x) = f(-x) = p(-x)q(-x)$ and $p(-x), q(-x) \in Z[x]$.

23. $(x + 3)(x + 5)(x + 6)$.

24. $(x + 1)^3$.

25. By the Mod 2 Irreducibility Test (Theorem 17.3 with $p = 2$) it is enough to show that $x^4 + x^3 + 1$ is irreducible over Z_2. By inspection, $x^4 + x^3 + 1$ has no zeros in Z_2 and so it has no linear factors over Z_2. The only quadratic irreducible in $Z_2[x]$ is $x^2 + x + 1$ and it is ruled out as a factor by long division.

26. For $f(x)$, both methods yield 4 and 5. (Notice that $\sqrt{-47} = \sqrt{2} = \pm 3$). Neither method yields a solution for $g(x)$. The quadratic formula applied to $g(x)$ involves $\sqrt{-23} = \sqrt{2}$ and there is no element of Z_5 whose square is 2. $ax^2 + bx + c$ $(a \neq 0)$ has a zero in $Z_p[x]$ if and only if $b^2 - 4ac = d^2$ for some d in Z_p.

27. **a.** Since every reducible polynomial of the form $x^2 + ax + b$ can be written in the form $(x - c)(x - d)$, we need only count the number of distinct such expressions over Z_p. Note that there are $p(p - 1)$ expressions of the form $(x - c)(x - d)$ where $c \neq d$. However, since $(x - c)(x - d) = (x - d)(x - c)$ there are only $p(p - 1)/2$ distinct such expressions. To these we must add the p cases of the form $(x - c)(x - c)$. This gives us $p(p - 1)/2 + p = p(p + 1)/2$.

b. First, note that for every reducible polynomial of the form $f(x) = x^2 + ax + b$ over Z_p the polynomial $cf(x)$ $(c \neq 0)$ is also reducible over Z_p. By part a, this gives us at least $(p - 1)p(p + 1)/2 = p(p^2 - 1)/2$ reducible polynomials over Z_p. Conversely, every quadratic polynomial over Z_p can be written in the form $cf(x)$ where $f(x)$ has lead coefficient 1. So, the $p(p^2 - 1)/2$ reducibles we have already counted include all cases.

28. Use Exercise 27.

29. By Exercise 28, for each prime p there is an irreducible polynomial $p(x)$ of degree 2 over Z_p. By Corollary 1 of Theorem 17.5, $Z_p[x]/\langle p(x)\rangle$ is a field. By the Division Algorithm (Theorem 16.2) every element in $Z_p[x]/\langle p(x)\rangle$ can be written in the form $ax + b + \langle p(x)\rangle$. Moreover, $ax + b + \langle p(x)\rangle = cx + d + \langle(p(x)\rangle$ only when $a = c$ and $c = d$ since $(ax + b) - (cx + d)$ is divisible by $p(x)$ only when it is 0. Thus, $Z_p[x]/\langle p(x)\rangle$ has order p^2.

30. By Eisenstein, $x^n + p$ where p is a prime is irreducible over Q.

31. Consider the mapping from $Z_3[x]$ onto $Z_3[i]$ given by $\phi(f(x)) = f(i)$. Since $\phi(f(x) + g(x)) = \phi((f + g)(x)) = (f + g)(i) = f(i) + g(i) = \phi(f(x)) + \phi(g(x))$ and $\phi(f(x)g(x)) = \phi((fg)(x) = (fg)(i) = f(i)g(i) = \phi(f(x))\phi(g(x))$, ϕ is a ring homomorphism. Because $\phi(x^2 + 1) = i^2 + 1 = -1 + 1 = 0$ we know that $x^2 + 1 \in \text{Ker } \phi$. From Theorem 16.4 we have that $\text{Ker } \phi = \langle x^2 + 1\rangle$. Finally, Theorem 15.3 gives us that $Z_3[x]/\langle x^2 + 1\rangle \approx Z_3[i]$.

32. If $a_2x^2 + a_1x + a_0 \in Z_p[x]$ is a factor of $f(x)$ with $a_2 \neq 0$ then $a_2^{-1}(a_2x^2 + a_1x + a_0)$ is a factor of $f(x)$.

33. $x^2 + 1, x^2 + x + 2, x^2 + 2x + 2$.

34. If so, then π is a zero of $x^2 - ax - b$.

35. We know that $a_n(r/s)^n + a_{n-1}(r/s)^{n-1} + \cdots + a_0 = 0$. So, clearing fractions we obtain $a_nr^n + sa_{n-1}r^{n-1} + \cdots + s^na_0 = 0$. This shows that $s \mid a_nr^n$ and $r \mid s^na_0$. By Euclid's Lemma (Chapter 0), s divides a_n or s divides r^n. Since s and r are relatively prime, s must divide a_n. Similarly, r must divide a_0.

36. Since $a_1(x)a_2(x)\cdots a_k(x) = a_1(x)(a_2(x)\cdots a_k(x))$, we have by Corollary 2 of Theorem 17.5 that $p(x)$ divides $a_1(x)$ or $p(x)$ divides $a_2(x)\cdots a_k(x)$. In the latter case, the Second Principle of Mathematical Induction implies that $p(x)$ divides some $a_i(x)$ for $i = 2, 3, \ldots, k$.

37. Suppose that $p(x)$ can be written in the form $g(x)h(x)$ where deg $g(x) <$ deg $p(x)$ and deg $h(x) <$deg $p(x)$ with $g(x),\ h(x) \in F[x]$. By Theorem 14.4, $F[x]/\langle p(x)\rangle$ is a field with
$0 + \langle p(x)\rangle = p(x) + \langle p(x)\rangle = g(x)h(x) + \langle p(x)\rangle = (g(x) + \langle p(x)\rangle)(h(x) + \langle p(x)\rangle)$.
Thus $g(x) + \langle p(x)\rangle = 0 + \langle p(x)\rangle$ or $h(x) + \langle p(x)\rangle = 0 + \langle p(x)\rangle$. This implies that $g(x) \in \langle p(x)\rangle$ or $h(x) \in \langle p(x)\rangle$. In either case we have contradicted Theorem 16.4.

38. Use the corollary to Theorem 17.4 and Exercise 5b with $a = -1$.

39. Since $(f + g)(a) = f(a) + g(a)$ and $(f \cdot g)(a) = f(a)g(a)$, the mapping is a homomorphism. Clearly, $p(x)$ belongs to the kernel. By Theorem 17.5, $\langle p(x)\rangle$ is a maximal ideal, so the kernel is $\langle p(x)\rangle$.

40. It follows from Theorem 17.1 that $\langle x^2 + 1\rangle$ is irreducible over Z and from Theorem 17.6 that if $f(x)g(x) \in \langle x^2 + 1\rangle$ where $f(x), g(x) \in Z[x]$ then $x^2 + 1$ is a factor of $f(x)$ or $g(x)$. So, $\langle x^2 + 1\rangle$ is a prime ideal.
For the second part, let $I = \langle x^2 + 1\rangle$ and observe that if I is a maximal ideal of $Z[x]$ then, by Theorem 14.4, $Z[x]/I = \{ax + b + I \mid a, b \in Z\}$ is field. Thus, the element $2 + I$ has a multiplicative inverse in $Z[x]/I$, say
$(2 + I)(ax + b) + I = 2(ax + b) + I = 1 + I$. Then, $2(ax + b) - 1$ is in I. Since every non-zero element in I has degree at least 2 and the left side is 0 or has degree at 1, we have that $2(ax + b) - 1 = 0$. But then the left side is odd.

41. Consider the mapping ϕ from F to $F[x]/\langle p(x)\rangle$ given by $\phi(a) = a + \langle p(x)\rangle$. By observation, ϕ is one-to-one and onto. Moreover,
$\phi(a + b) = a + b + \langle p(x)\rangle = a + \langle p(x)\rangle + b + \langle p(x)\rangle = \phi(a) + \phi(b)$ and
$\phi(ab) = ab + \langle p(x)\rangle = (a + \langle p(x)\rangle)(b + \langle p(x)\rangle) = \phi(a)\phi(b)$ so ϕ is a ring isomorphism.

42. Let $f(x) = g(x)h(x)$ where deg $g(x) <$ deg $f(x)$ and deg $h(x) <$ deg $f(x)$. Then $g(x)h(x)$ belongs to $\langle f(x)\rangle$ but neither $g(x)$ nor $h(x)$ belongs to $\langle f(x)\rangle$.

43. $f(x)$ is primitive.

44. 11

45. Since $x^2 + x + 1$ is the only quadratic irreducible over Z_2, the possibilities are $x^2(x^2 + x + 1), (x + 1)^2(x^2 + x + 1)$, and $x(x + 1)(x^2 + x + 1)$.

46. Note that by Exercise 49 in Chapter 13, $(x + 2)^3 = (x - 1)^3$.

47. Let $f(x) = (5/7)x^4 + (9/2)x^3 + (3/4)x^2 + 6x + 1/2$. Then, clearing fractions, we have $28f(x) = 20x^4 + 14 \cdot 9x^3 + 21x^2 + 28 \cdot 6x + 14$. This polynomial meets the conditions for Eisenstein's Criterion with the prime 7. So, $28f(x)$ is irreducible over Q. But then by Exercise 5a, $f(x)$ is also irreducible over Q.

48. By Eisenstein, $p(x) = x^5 + 15x^3 - 20x - 10$ is irreducible over Z. Suppose that $g(x)h(x)$ is in $\langle p(x)\rangle$. Then $g(x)h(x) = p(x)k(x)$ for some $k(x)$ in $Z[x]$. By the unique factorization over Z property, we have that $p(x)$ must be one of the irreducible factors of $g(x)h(x)$. So, one of $g(x)$ or $h(x)$ has $p(x)$ as a factor and therefore, one of them is in $\langle p(x)\rangle$.

49. If $a \in Z_p$ is a zero of $x^2 + x + 1$, then $a^3 = 1$. So, $|a| = 3$, and therefore 3 divides $|U(p)| = p - 1$. This gives us that 3 divides both $p - 1$ and $p - 2$, But then 3 must divide $(p - 1) - (p - 2) = 1$.

50. By Theorem 17.3, we can say that $f(x)$ is irreducible over Q, provided that the lead coefficient is not divisible by p.

51. The general result is that $\langle x, n \rangle$ is prime ideal in $Z[x]$ if and only if $n = 0$ or $|n|$ is a prime.

    To see this note that $I = \langle x, n \rangle = \{f(x) \in Z[x] \mid f(0) = n\}$. If $n = ab$ where $1 < a < n$, $g(x) = a$ and $h(x) = b$ have the property that $g(x)h(x)$ belongs to I but $g(x)$ and $h(x)$ do not.

52. The degree is at least 3 or $f(x) - g(x) = 0$, and so $f(x) - g(x)$ has no degree.

CHAPTER 18

Divisibility in Integral Domains

1. 1. $|a^2 - db^2| = 0$ implies $a^2 = db^2$. Thus $a = 0 = b$, since otherwise $d = 1$ or d is divisible by the square of a prime.
   2. $N((a + b\sqrt{d})(a' + b'\sqrt{d})) = N(aa' + dbb' + (ab' + a'b)\sqrt{d}) =$ $|(a^2 - db^2)(a'^2 - db'^2)| = |(aa' + dbb')^2 - d(ab' + a'b)^2| =$ $|a^2a'^2 + d^2b^2b'^2 - da^2b'^2 - da'^2b^2| = |a^2 - db^2||a'^2 - db'^2| = N(a + b\sqrt{d})N(a' + b'\sqrt{d})$.
   3. If $xy = 1$, then $1 = N(1) = N(xy) = N(x)N(y)$ and $N(x) = 1 = N(y)$. If $N(a + b\sqrt{d}) = 1$, then $\pm 1 = a^2 - db^2 = (a + b\sqrt{d})(a - b\sqrt{d})$ and $a + b\sqrt{d}$ is a unit.
   4. This part follows directly from 2 and 3.

2. Say $a = bu$ where u is a unit. Then $ra = rbu = (ru)b \in \langle b \rangle$ so that $\langle a \rangle \subseteq \langle b \rangle$. By symmetry, $\langle b \rangle \subseteq \langle a \rangle$. If $\langle a \rangle = \langle b \rangle$, then $a = bu$ and $b = av$. Thus, $a = avu$ and $uv = 1$.

3. Let $I = \cup I_i$. Let $a, b \in I$ and $r \in R$. Then $a \in I_i$ for some i and $b \in I_j$ for some j. Thus $a, b \in I_k$, where $k = \max\{i, j\}$. So, $a - b \in I_k \subseteq I$ and ra and $ar \in I_k \subseteq I$.

4. Say r is irreducible and u is a unit. If $ru = ab$ where a and b are not units, then $r = a(bu^{-1})$ where a and bu^{-1} are not units.

5. Clearly, $\langle ab \rangle \subseteq \langle b \rangle$. So the statement is equivalent to $\langle ab \rangle = \langle b \rangle$ if and only if a is a unit. If $\langle ab \rangle = \langle b \rangle$ there is an r in the domain such that $b = rab$, so that $1 = ra$ and a is a unit. If a is a unit then $b = a^{-1}(ab)$ belongs to $\langle ab \rangle$ and therefore $\langle b \rangle \subseteq \langle ab \rangle$.

6. $a \sim a$ since $a = a \cdot 1$; if $a \sim b$, say $a = bu$ where u is a unit, then $b = au^{-1}$ so $b \sim a$; if $a \sim b$, say $a = bu$ where u is a unit and $b \sim c$, say $b = cr$ where r is a unit, then $a = bu = cru$ where ru is a unit.

   $\langle ab \rangle = \langle b \rangle$. So, a is not a unit.

7. Say $x = a + bi$ and $y = c + di$. Then

$$xy = (ac - bd) + (bc + ad)i.$$

   So

$$d(xy) = (ac - bd)^2 + (bc + ad)^2 = (ac)^2 + (bd)^2 + (bc)^2 + (ad)^2.$$

   On the other hand,

$$d(x)d(y) = (a^2 + b^2)(c^2 + d^2) = a^2c^2 + b^2d^2 + b^2c^2 + a^2d^2.$$

8. First, observe that for any $r \in D, d(1) \leq d(1 \cdot r) = d(r)$ so that $d(1)$ is the minimum value of d. Now if u is a unit, then $d(u) \leq d(uu^{-1}) = d(1)$ so that $d(u) = d(1)$. If $d(u) = d(1)$, then $1 = uq + r$ where $r = 0$ or $d(r) < d(u) = d(1)$. So $r = 0$.

9. Suppose $a = bu$, where u is a unit. Then $d(b) \leq d(bu) = d(a)$. Also, $d(a) \leq d(au^{-1}) = d(b)$.

10. Mimic the proof of Theorem 17.5.

11. Suppose that $x = a + b\sqrt{d}$ is a unit in $Z[\sqrt{d}]$. Then $1 = N(x) = a^2 + (-d)b^2$. But $-d > 1$ implies that $b = 0$ and $a = \pm 1$.

12. Suppose that D has a proper ideal I_1 that is not contained in a maximal ideal. Let a_1 be in D but not in I_1. Then $I_2 = \langle I_1, a_1 \rangle$ is a proper ideal that properly contains I_1 and is not maximal. Repeating this argument we have a strictly increasing chain of ideals $I_1 \subset I_2 \subset \cdots$. So, by the Ascending Chain Condition, this chain is finite. But then the last ideal in the chain is maximal.
Alternate solution. Suppose that D has a proper ideal I_1 that is not contained in a maximal ideal. By definition, there is some proper ideal I_2 that properly contains I_1 but is not a maximal ideal. Repeating this argument we have a strictly increasing chain of ideals $I_1 \subset I_2 \subset \cdots$. So, by the Ascending Chain Condition, this chain is finite. But then the last ideal in the chain is maximal.

13. First, observe that $21 = 3 \cdot 7$ and that $21 = (1 + 2\sqrt{-5})(1 - 2\sqrt{-5})$. To prove that 3 is irreducible in $Z[\sqrt{-5}]$ suppose that $3 = xy$, where $x, y \in Z[\sqrt{-5}]$ and x and y are not units. Then $9 = N(3) = N(x)N(y)$ and, therefore, $N(x) = N(y) = 3$. But there are no integers a and b such that $a^2 + 5b^2 = 3$. The same argument shows that 7 is irreducible over $Z[\sqrt{-5}]$. To show that $1 + 2\sqrt{-5}$ is irreducible over $Z[\sqrt{-5}]$ suppose that $1 + 2\sqrt{-5} = xy$, where $x, y \in Z[\sqrt{-5}]$ and x and y are not units. Then $21 = N(1 + 2\sqrt{-5}) = N(x)N(y)$. Thus $N(x) = 3$ or $N(x) = 7$, both of which are impossible.

14. Suppose that there are Gaussian integers z_1 and z_2 such that $1 - i = z_1 z_2$. Then $2 = d(1 - i) = d(z_1)d(z_2)$. This forces $d(z_1) = 1$ or $d(z_2) = 1$. Now use Exercise 8.

15. First, observe that $10 = 2 \cdot 5$ and that $10 = (2 - \sqrt{-6})(2 + \sqrt{-6})$. To see that 2 is irreducible over $Z[\sqrt{-6}]$, assume that $2 = xy$, where $x, y \in Z[\sqrt{-6}]$ and x and y are not units. Then $4 = N(2) = N(x)N(y)$ so that $N(x) = 2$. But 2 cannot be written in the form $a + 6b^2$. A similar argument applies to 5. To see that $2 - \sqrt{-6}$ is irreducible, suppose that $2 - \sqrt{-6} = xy$ where $x, y \in Z[\sqrt{-6}]$ and x and y are not units. Then $10 = N(2 - \sqrt{-6}) = N(x)N(y)$ and as before, this is impossible. We know that $Z[\sqrt{-6}]$ is not a principle ideal domain because a PID is a UFD (Theorem 18.3).

16. $\mathbf{C}[x]$ is a UFD but contains $Z[\sqrt{-6}]$.

17. Suppose $3 = \alpha\beta$, where $\alpha, \beta \in Z[i]$ and neither is a unit. Then $9 = d(3) = d(\alpha)d(\beta)$, so that $d(\alpha) = 3$. But there are no integers such that $a^2 + b^2 = 3$. Observe that $2 = -i(1 + i)^2$ and $5 = (1 + 2i)(1 - 2i)$ and $1 + i, 1 + 2i$, and $1 - 2i$ are not units.

18. If $7 = (a + b\sqrt{6})(c + d\sqrt{6})$ and neither factor is a unit, then $|a^2 - 6b^2| = 7$ and, modulo 7, $a^2 = 6b^2$. However the only solutions to this equation modulo 7 are $a = b = 0$. In this case $c + d\sqrt{6}$ is a unit.

19. Use Exercise 1 with $d = -1$. 5 and $1 + 2i$; 13 and $3 + 2i$; 17 and $4 + i$.

20. Use Example 1 and Theorem 18.2.

21. Suppose that $1 + 3\sqrt{-5} = xy$, where $x, y \in Z[\sqrt{-5}]$ and x and y are not units. Then $46 = N(1 + 3\sqrt{-5}) = N(x)N(y)$. Thus, $N(x) = 2$ or $N(x) = 23$. But neither 2 nor 5 can be written in the form $a^2 + 5b^2$, so $1 + 3\sqrt{-5}$ is irreducible over $Z[\sqrt{-5}]$. To see that $1 + 3\sqrt{-5}$ is not prime, observe that $(1 + 3\sqrt{-5})(1 - 3\sqrt{-5}) = 1 + 45 = 46$ so that $1 + 3\sqrt{-5}$ divides $2 \cdot 23$. For $1 + 3\sqrt{-5}$ to divide 2, we need $46 = N(1 + 3\sqrt{-5})$

divides $N(2) = 4$. Likewise, for $1 + 3\sqrt{-5}$ to divide 23 we need that 46 divides 23^2. Since neither of these is true, $1 + 3\sqrt{-5}$ is not prime.

22. To show the two elements are irreducible mimic the solution given for Exercise 18 but use modulo 4. To show the two elements are not prime use the observation that $2 \cdot 2 = 4 = (1 + \sqrt{5})(-1 + \sqrt{5})$. Also note that 2 does not divide $(1 + \sqrt{5})$ for if so, there would be an element z in $Z[\sqrt{5}]$ such that $2z = 1 + \sqrt{5}$. But $N(2z) = N(2)N(z) = 4N(z) = N(1 + \sqrt{5}) = 4$. Thus $N(z) = 1$. But the only elements of $Z[\sqrt{5}]$ with norm 1 are 1 and -1. A similar argument shows that 2 does not divide $-1 = \sqrt{5}$ and, neither $1 + \sqrt{5}$ nor $-1 + \sqrt{5}$ divides 2.

23. First, observe that $(-1 + \sqrt{5})(1 + \sqrt{5}) = 4 = 2 \cdot 2$ and by Exercise 22, $1 + \sqrt{5}$ and 2 are irreducible over $Z[\sqrt{5}]$. To see that $-1 + \sqrt{5}$ is irreducible over $Z[\sqrt{5}]$, suppose that $-1 + \sqrt{5} = xy$ where $x, y \in Z[\sqrt{5}]$ and x and y are not units. Let $x = a + b\sqrt{5}$. Then $4 = N(-1 + \sqrt{5}) = N(x)N(y)$ so that $a^2 - 5b^2 = \pm 2$. Viewing this equation modulo 5 gives us $a^2 = 2$ or $a^2 = -2 = 3$. However, every square in Z_5 is 0, 1, or 4.

24. Let I be a non-zero prime ideal in $F[x]$ and let M be a proper ideal that contains I. By Theorem 16.3 there are elements $f(x)$ and $g(x)$ in $F[x]$ such that $M = \langle f(x) \rangle$ and $I = \langle g(x) \rangle$. By definition $g(x)$ is a prime and by Theorem 18.2 it is irreducible over F. Since M contains I there is an $h(x)$ in $F[x]$ such that $g(x) = f(x)h(x)$ and since $g(x)$ is irreducible over F, $h(x)$ is a unit. Thus by Exercise 2 in this chapter $M = I$.

25. $m = 0$ and $n = -1$ give $q = -i, r = -2 - 2i$.

26. Observe that $(3 + 2\sqrt{2})(3 - 2\sqrt{2}) = 1$. So, $(3 + 2\sqrt{2})^n(3 - 2\sqrt{2})^n = ((3 + 2\sqrt{2})(3 - 2\sqrt{2}))^n = 1$.

27. $1 = N(ab) = N(a)N(b)$ so $N(a) = 1 = N(b)$.

28. To see that 2 is prime in Z_{12} note that 2 is not a unit in Z_{12} and suppose that $bc = 2t$ where b, c and t belong to Z_{12}. Thus there is an integer k such that $bc - 2t = 12k$ in the ring Z. This implies that 2 divides bc in Z. Thus one of b or c is divisible by 2 in Z_{12}. The same argument applies when 2 is replaced by 3. To see that 2 is irreducible in Z_{12}, suppose that $2 = bc$ in Z_{12}. Then there is an integer k such that $2 - bc = 12k$ in Z. By Euclid's Lemma this implies that 2 divides b or c in Z. Say, $b = 2d$. Then in Z we have $2 - 2dc = 12k$ which reduces to $1 = dc + 6k$. This implies that c is relatively prime to 6 and therefore a unit in Z_{12}. But the only solutions to the equation $2 = 2x$ in Z_{12} are $x = 1$ and $x = 7$. If $dc = 1$ then c is a unit and if $dc = 7$ then in Z_{12}, $7dc = 7 \cdot 7 \bmod 12 = 1$ and c is again a unit. To see that 3 is reducible in Z_{12} note that $3 = 3 \cdot 9 \bmod 12$ and neither 3 nor 9 is a unit.

29. Suppose that $bc = pt$ in Z_n. Then there exists an integer k such that $bc = pt + kn$. This implies that p divides bc in Z and by Euclid's Lemma we know that p divides b or p divides c.

30. If p^2 does not divide n then there are integers s and t such that $1 = ps + (n/p)t$. Thus, $p = p(ps) \bmod n$. Since both p and ps are zero-divisors in Z_n they cannot be units. This contradicts our assumption that p is irreducible in Z_n.

31. See Example 3.

32. If $(a + bi)$ is a unit, then $a^2 + b^2 = 1$. Thus, $\pm 1, \pm i$.

33. Note that $p|(a_1a_2 \cdots a_{n-1})a_n$ implies that $p|a_1a_2 \cdots a_{n-1}$ or $p|a_n$. Thus, by induction, p divides some a_i.

34. $4x + 1$ and $2x + 3$ are associates of $3x + 2$ and $x + 4$.

35. By Exercise 10, $\langle p \rangle$ is maximal and by Theorem 14.4, $D/\langle p \rangle$ is a field.

36. Let a be a nonzero element of the domain. It suffices to show that a is a unit. Consider the chain $\langle a \rangle \supseteq \langle a^2 \rangle \supseteq \langle a^3 \rangle \supseteq \langle a^4 \rangle \supseteq \cdots$. By hypothesis, we have $\langle a^n \rangle = \langle a^{n+1} \rangle$ for some n. Thus $a^n = a^{n+1}b$ for some b and $1 = ab$.

37. Suppose R satisfies the ascending chain condition and there is an ideal I of R that is not finitely generated. Then pick $a_1 \in I$. Since I is not finitely generated, $\langle a_1 \rangle$ is a proper subset of I, so we may choose $a_2 \in I$ but $a_2 \notin \langle a_1 \rangle$. As before, $\langle a_1, a_2 \rangle$ is proper, so we may choose $a_3 \in I$ but $a_3 \notin \langle a_1, a_2 \rangle$. Continuing in this fashion, we obtain a chain of infinite length $\langle a_1 \rangle \subset \langle a_1, a_2 \rangle \subset \langle a_1, a_2, a_3 \rangle \subset \cdots$.

    Now suppose every ideal of R is finitely generated and there is a chain $I_1 \subset I_2 \subset I_3 \subset \cdots$. Let $I = \cup I_i$. Then $I = \langle a_1, a_2, \ldots, a_n \rangle$ for some choice of $a_1, a_2, \ldots, a_n$. Since $I = \cup I_i$, each a_i belongs to some member of the union, say $I_{i'}$. Letting $k = \max\{i' \mid i = 1, \ldots, n\}$, we see that all $a_i \in I_k$. Thus, $I \subseteq I_k$ and the chain has length at most k.

38. The set of complex numbers is a Euclidean domain (take $d(a) = 0$ for all $a \neq 0$) containing $Z[\sqrt{-5}]$ as a subdomain. Now use Example 7 and the corollary to Theorem 18.4.

39. Say $I = \langle a + bi \rangle$. Then $a^2 + b^2 + I = (a + bi)(a - bi) + I = I$ and therefore $a^2 + b^2 \in I$. For any $c, d \in Z$, let $c = q_1(a^2 + b^2) + r_1$ and $d = q_2(a^2 + b^2) + r_2$, where $0 \leq r_1, r_2 < a^2 + b^2$. Then $c + di + I = r_1 + r_2 i + I$.

40. $-1 + \sqrt{2}$; infinite.

41. $N(6 + 2\sqrt{-7}) = 64 = N(1 + 3\sqrt{-7})$. The other part follows directly from Exercise 25.

42. Let $I_i = \{(a_1, a_2, \ldots, a_i, 0, 0, 0, \ldots)\}$, where the a's are integers.

43. Theorem 18.1 shows that primes are irreducible. So, assume that a is an irreducible in a UFD R and that $a|bc$ in R. We must show that $a|b$ or $a|c$. Since $a|bc$, there is an element d in R such that $bc = ad$. Now replacing b, c, and d by their factorizations as a product of irreducibles, we have by the uniqueness property that a (or an associate of a) is one of the irreducibles in the factorization of bc. Thus, a is a factor of b or a is a factor of c.

44. Observe that both x^2 and x^3 are irreducible over F but $x^3x^3 = x^2x^2x^2$.

45. See Exercise 23 in Chapter 0.

46. The argument given in Example 7 of Chapter 18 applies in both cases.

CHAPTER 19

Extension Fields

1. $\{a5^{2/3} + b5^{1/3} + c \mid a, b, c \in Q\}$.

2. Observe that $\frac{1}{\sqrt{2}+\sqrt{3}} = \frac{1}{\sqrt{2}+\sqrt{3}}\frac{\sqrt{2}-\sqrt{3}}{\sqrt{2}-\sqrt{3}} = -\sqrt{2}+\sqrt{3}$ is in $Q(\sqrt{2}+\sqrt{3})$. So, $(\sqrt{2}+\sqrt{3}+(-\sqrt{2}+\sqrt{3}) = 2\sqrt{3}$ is in $Q(\sqrt{2}+\sqrt{3})$. Then, $\sqrt{3}$ and $(\sqrt{2}+\sqrt{3}) - \sqrt{3} = \sqrt{2}$ are in $Q(\sqrt{2}+\sqrt{3})$.

   Alternate solution. Observe that $(\sqrt{2}+\sqrt{3})^2 = 5 + 2\sqrt{2}\sqrt{3}$ is in $Q(\sqrt{2}+\sqrt{3})$. From this we have that $\sqrt{2}\sqrt{3}$ is in $Q(\sqrt{2}+\sqrt{3})$. Then, $(\sqrt{2}\sqrt{3})(\sqrt{2}+\sqrt{3}) = 2\sqrt{3}+3\sqrt{2} = 3\sqrt{2}+2\sqrt{3}$ is in $Q(\sqrt{2}+\sqrt{3})$. Moreover, $3(\sqrt{2}+\sqrt{3}) - (3\sqrt{2}+2\sqrt{3}) = \sqrt{3}$ is in $Q(\sqrt{2}+\sqrt{3})$. Then, $(\sqrt{2}+\sqrt{3}) - \sqrt{3} = \sqrt{3}$ is in $Q(\sqrt{2}+\sqrt{3})$. Finally, $\sqrt{2}+\sqrt{3}$ is in $Q(\sqrt{2}, \sqrt{3})$ by closure.

3. Since $x^3 - 1 = (x-1)(x^2+x+1)$, the zeros of $x^3 - 1$ are $1, (-1+\sqrt{-3})/2$, and $(-1-\sqrt{-3})/2$. So, the splitting field is $Q(\sqrt{-3})$.

4. $F(a,b) \subseteq F(a)(b)$ since $F(a)(b)$ is a field that contains F, a and b. Also, $F(a)(b) \subseteq F(a,b)$ since $F(a,b)$ contains $F(a)$ and b. So, $F(a,b) = F(a)(b)$ and, by symmetry, $F(a,b) = F(b,a)$.

5. Since the zeros of x^2+x+1 are $(-1 \pm \sqrt{-3})/2$ and the zeros of $x^2 - x + 1$ are $(1 \pm \sqrt{-3})/2$, the splitting field is $Q(\sqrt{-3})$.

6. $2^{2/3} = (2^{1/3})^2 \in Q(2^{1/3})$ by closure. So, $Q(2^{2/3}) \subseteq Q(2^{1/3})$. By closure, $2 \cdot 2^{1/3} = (2^{2/3})^2 \in Q(2^{2/3})$. So, $Q(2^{1/3}) \subseteq Q(2^{2/3})$.

7. Since $ac + b \in F(c)$, we have $F(ac+b) \subseteq F(c)$. Also, $c = a^{-1}(ac+b) - a^{-1}b$, so $F(c) \subseteq F(ac+b)$.

8. 8. Use Theorem 19.3. To construct the multiplication table, observe that $a^3 = a+1$.

9. Since $a^3+a+1 = 0$, we have $a^3 = a+1$. Thus, $a^4 = a^2+a; a^5 = a^3+a^2 = a^2+a+1$. To compute a^{-2} and a^{100}, we observe that $a^7 = 1$, since $F(a)^*$ is a group of order 7. Thus, $a^{-2} = a^5 = a^2+a+1$ and $a^{100} = (a^7)^{14}a^2 = a^2$.

10. Use the fact that $a^3 = a+1$, $a^4 = a^2+a$, $a^5 = a^2+a+1$, $a^6 = a^2+1$ and $a^7 = 1$.

11. $Q(\pi)$ is the set of all expressions of the form

    $$(a_n\pi^n + a_{n-1}\pi^{n-1} + \cdots + a_0)/(b_m\pi^m + b_{m-1}\pi^{m-1} + \cdots + b_0),$$

    where $b_m \neq 0$.

12. A basis is $\{1, \pi, \pi^2\}$. $1/\pi = (1/\pi^3)\pi^2$; $\pi^{11} = ((\pi^3)^3)\pi^2$.

13. $x^7 - x = x(x^6-1) = x(x^3+1)(x^3-1) = x(x-1)^3(x+1)^3; x^{10} - x = x(x^9-1) = x(x-1)^9$ (see Exercise 49 of Chapter 13).

14. Suppose that ϕ is an automorphism of $Q(\sqrt{5})$. Since $\phi(1) = 1$, we have $\phi(n) = \phi(n \cdot 1) = n\phi(1) = n$. Also, $1 = \phi(n/n) = n\phi(1/n)$ gives $\phi(1/n) = 1/n$. Thus, $\phi(m/n) = m\phi(1/n) = m/n$. So ϕ is the identity map on Q. Lastly, $5 = \phi(5) = \phi(\sqrt{5}^2) = (\phi(\sqrt{5}))^2$, so $\phi(\sqrt{5}) = \pm\sqrt{5}$. So there are two automorphisms of $Q(\sqrt{5})$. For the case of $Q(\sqrt[3]{5})$ we have that ϕ is the identity map on Q and $5 = \phi(5) = \phi(\sqrt[3]{5}^3) = (\phi(\sqrt[3]{5}))^3$, so $\phi(\sqrt[3]{5}) = \sqrt[3]{5}$. So there is only the identity automorphism of $Q(\sqrt[3]{5})$.

15. If $f(x)$ is irreducible over F we are done. If $f(x)$ does not split in F, then $f(x) = g(x)h(x)$ where $g(x), h(x) \in F[x]$, $1 \leq \deg g(x) < p$, and $g(x)$ is irreducible over F. Let b be a zero of $f(x)$ in some extension of F. Then $b^p = a$ and $f(x) = x^p - b^p = (x - b)^p$ (see Exercise 49 of Chapter 13). If $b \in F$, then $f(x)$ splits in F; if $b \notin F$, then $\deg g(x) > 1$ and has multiple zeros. So, by Theorem 19.6, we know that $g(x) = k(x^p)$ for some $k(x)$ in $F[x]$. But then $g(x)$ has degree at least p. Alternate proof. Let $g(x)$ and b be as in the first proof. Because $f(x) = (x - b)^p$ we know that $g(x) = (x - p)^k$ for some $1 < k < p$. In the expanded product, the coefficient of x^{k-1} is kb and since $g(x) \in F[x]$ we have that $kb \in F$. Since $k < p$, we know that k is in F. Then $k^{-1}kb = b$ is in F and $f(x)$ splits in F.

16. By Theorem 19.6, we obtain $(x + \beta)(x + \beta^2)(x + \beta^4)(x + \beta^8) = (x + \beta)(x + \beta^2)(x + \beta + 1)(x + \beta^2 + 1)$.

17. Solving $1 + \sqrt[3]{4} = (a + b\sqrt[3]{2} + c\sqrt[3]{4})(2 - 2\sqrt[3]{2})$ for a, b, and c yields $a = 4/3, b = 2/3$, and $c = 5/6$.

18. $a = -3/23, b = 4/23$.

19. By Exercise 7, $Q(4 - i) = Q$ and $Q(1 + i) = Q$.
Alternate solution. Since $1 + i = -(4 - i) + 5$, $Q(1 + i) \subseteq Q(4 - i)$; conversely, $4 - i = 5 - (1 + i)$ implies that $Q(4 - i) \subseteq Q(1 + i)$.

20. Note that $a = \sqrt{1 + \sqrt{5}}$ implies that $a^4 - 2a^2 - 4 = 0$. Then $p(x) = x^4 - 2x^2 - 4$ is irreducible over Q. To see this, use the mod 3 test on $x^4 - 2x^2 + 2$ Substitution shows this has no zeros. By Example 8 in Chapter 17, the only quadratic factors we need check as factors are $x^2 + 1, x^2 + x + 2$ and $x^2 = 2x + 2$. Long division rules these out.

21. If the zeros of $f(x)$ are $a_1, a_2, \ldots, a_n$, then the zeros of $f(x + a)$ are $a_1 - a, a_2 - a, \ldots, a_n - a$. So, by Exercise 7, $f(x)$ and $f(x - a)$ have the same splitting field.

22. Since $f(x)$ and $g(x)$ are relatively prime in $F[x]$ there are polynomials $t(x)$ and $s(x)$ in $F[x]$ such that $1 = f(x)t(x) + g(x)s(x)$. (See Exercise 43 for Chapter 16.) If $f(x)$ and $g(x)$ had a nonconstant factor in common this factor would divide 1.

23. Clearly, Q and $Q(\sqrt{2})$ are subfields of $Q(\sqrt{2})$. Assume that there is a subfield F of $Q(\sqrt{2})$ that contains an element $a + b\sqrt{2}$ with $b \neq 0$. Then, since every subfield of $Q(\sqrt{2})$ must contain Q, we have by Exercise 20 that $Q(\sqrt{2}) = Q(a + b\sqrt{2}) \subseteq F$. So, $F = Q(\sqrt{2})$.

24. They are of the form $a + b\sqrt[4]{2}$ where $a, b \in Q(\sqrt{2})$.

25. It is 64. To see this, we let a be a zero in the splitting field of $x^2 + x + 1$ over Z_2. Then $x^2 + x + 1 = (x - a)(x - a - 1)$. Checking to see that none of the four elements of $F(a)$ is a zero of $x^3 + x + 1$, we know that $x^3 + x + 1$ is irreducible over $F(a)$.

Then letting b be a zero of $x^3 + x + 1$ in the splitting field, we know that $F(a)(b)$ is a field of order 64 in which $x^3 + x + 1$ splits.

26. Following Example 8, we first observe that i is a primitive 4th root of unity. Then the splitting field is $Q(\sqrt[4]{-1}, i) = Q(\sqrt[4]{-1})$ since $(\sqrt[4]{-1})^2 = i$.

27. Let a be a zero of $x^3 + 2x + 1$ in some extension of Z_3. By Theorem 19.3, $F(a) = \{c_2a^2 + c_1a + c_0 \mid \text{where } c_2, c_1, c_0 \in Z_3\}$ and we know that $a^3 + 2a + 1 = 0$. Using long division, we obtain $x^3 + 2x + 1 = (x - a)(x^2 + ax + (2 - a^2))$. By trial and error, we discover that $a + 1$ is a zero of $x^2 + ax + (2 - a^2)$ and, by long division, we deduce that $-2a - 1$ is the other zero of $x^2 + ax + (2 - a^2)$. So, we have $x^3 + 2x + 1 = (x - a)(x - a - 1)(x + 2a + 1)$.

28. Over Z_2, $x^6 + 1 = x^6 - 1 = (x^3 - 1)(x^3 + 1) = (x^3 - 1)^2 = (x - 1)^2(x^2 + x + 1)^2 = (x + 1)^2(x^2 + x + 1)^2$.
Over Z_3, we use Exercise 49 in Chapter 13 to obtain $x^6 + 1 = (x^2)^3 + 1^3 = (x^2 + 1)^3$.

29. Suppose that $\phi\colon Q(\sqrt{-3}) \to Q(\sqrt{3})$ is an isomorphism. Since $\phi(1) = 1$, we have $\phi(-3) = -3$. Then $-3 = \phi(-3) = \phi(\sqrt{-3}\sqrt{-3}) = (\phi(\sqrt{-3}))^2$. This is impossible, since $\phi(\sqrt{-3})$ is a real number.

30. The field of quotients of $Z_p[x]$ is not perfect.

31. By long division we obtain $x^2 + x + 2 = (x - \beta)(x + \beta + 1)$, so the other zero is $-\beta - 1 = 4\beta + 4$.

32. Use Theorem 19.5.

33. Since $f(x) = x^{21} + 2x^8 + 1$ and $f'(x) = x^7$ have no common factor of positive degree, we know by Theorem 19.5 that $f(x)$ has no multiple zeros in any extension of Z_3.

34. Observe that because 1 is a zero of both $f(x) = x^{19} + x^8 + 1$ and $f'(x) = 19x^{18} + 8x^7 = x^{18} + 2x^7$ we know by Corollary 2 of Theorem 16.2 (the Factor Theorem) and Theorem 19.5 that $f(x)$ has multiple zeros in some extension of Z_3. By the proof on Theorem 19.5, 1 is a multiple zero.

35. Since $f(x) = x^{p^n} - x$ and $f'(x) = -1$ have no common factor of positive degree, we know by Theorem 19.5 that $f(x)$ has no multiple zeros in any extension of Z_3.

36. The splitting field is $F = Z_3[x]/\langle x^2 + x + 2\rangle$ and β is a zero of $x^2 + x + 2$ in F. F has nine elements and $f(x) = (x - \beta)(x - (2\beta + 2))(x - 2\beta)(x - (\beta + 1))$.

37. Let K be the intersection of all the subfields of E that contain F and the set $\{a_1, a_2, \ldots, a_n\}$. It follows from the subfield test given in Exercise 29 Chapter 13 that K is a subfield of E and, by definition, K contains F and the set $\{a_1, a_2, \ldots, a_n\}$. Since $F(a_1, a_2, \ldots, a_n)$ is the smallest such field, we have $F(a_1, a_2, \ldots, a_n) \subseteq K$. Moreover, since the field $F(a_1, a_2, \ldots, a_n)$ is one member of the intersection, we have $K \subseteq F(a_1, a_2, \ldots, a_n) \subseteq K$.

38. Observe that $x^4 - x^2 - 2 = x^4 + 2x^2 + 1 = (x^2 + 1)^2$. So the splitting field is $Z_3[x]/\langle x^2 + 1\rangle$.

39. Since $|(Z_2[x]/\langle f(x)\rangle)^*| = 31$ is prime and the order of every element must divide it, every nonidentity is a generator.

40. Observe that $x^4 - 6x^2 - 7 = (x^2 - 7)(x^2 + 1)$.

41. Use the Fundamental Theorem of Field Theory (Theorem 19.1) and the Factor Theorem (Corollary 2 of Theorem 16.2).

42. By the proof of the Fundamental Theorem of Field Theory $p(x)$ has a zero in some extension E of F and $[E : F] \leq n$. Now write $p(x) = (x - a)g(x)$ where $g(x) \in E[x]$. By induction on the degree of the polynomial, $g(x)$ splits in some extension K of E and $[K : E] \leq (n-1)!$. Thus $p(x)$ splits in K and $[K : F] = [K : E][E : F] \leq (n-1)!n = n!$.

43. Proceeding as in Example 9 we suppose that $h(t)/k(t)$ is a zero in $Z_p(t)$ of $f(x)$ where deg $h(t) = m$ and deg $k(t) = n$. Then $(h(t)/k(t))^p = t$, and therefore $(h(t))^p = t(k(t))^p$. Then by Exercise 49 of Chapter 13 we have $h(t^p) = tk(t^p)$. Since deg $h(t^p) = pm$ and deg $tk(t^p) = 1 + pn$, we have $pm = 1 + pn$. But this implies that p divides 1, which is false. So, our assumption that $f(x)$ has a zero in $Z_p(x)$ has led to a contradiction. That $f(x)$ has a multiple zero in K follows as in Example 9.

44. By the corollary to Theorem 19.9, deg $f(x)$ has the form nt where t is the number of distinct zeros of $f(x)$.

45. Since $-1 = 1$, $x^n - x$ would have 1 as a multiple zero. But then, by Theorem 19.5, $x^n - x$ and its derivative, which is $-1 = 1$, must have a common factor of positive degree. This is impossible.

46. Since $x^2 + x + 1$ is the only irreducible over F, it is the only possible quadratic factor in the product. If $x^2 + x + 1$ appeared more than once in the product, then in the irreducible factorization of $f(x)$ in the splitting field of $f(x)$ over F the linear factors of $f(x)$ would have multiplicity at least 2, in violation of the corollary of Theorem 19.9.

47. From Example 8 we know that the splitting field of $x^3 - 2$ over Q is $Q(\sqrt[3]{2}, \omega)$ where $\omega = -1/2 + \sqrt{3}i/2$. So, the splitting field over $F = Q(\sqrt[3]{2})$ is $F(\omega)$ where $\omega = -1/2 + \sqrt{3}i/2$. The splitting field over $F = Q(\sqrt{3}i)$ is $F(\sqrt[3]{2})$.

48. $2a^2 + 1$.

49. Observe that the polynomial $x^2 - 2x - 1$ is irreducible over Z_5. So Theorems 19.1 and 19.3 shows that such a field exists.

50. No, because $\alpha^2 = \alpha + 2$ implies that $0 = \alpha^2 - \alpha - 2 = (\alpha + 1)(\alpha - 2)$ and neither $\alpha + 1$ nor $\alpha - 2$ is 0.

51. If $\alpha \in F(\beta)$, then we have $\alpha = a\beta + b$ for some $a, b \in F$. Squaring both sides, replacing β^2 with $-\beta - 1$, and solving for β, we find that $\beta \in F$. For the second part, if $\beta \in F(\alpha)$ we have $\beta = a\alpha + b$ for some $a, b \in F$. Solving for α in terms of β and proceeding as before, we get that β is in F.

52. $F_0 = Q(i), F_1 = Q(i, \sqrt{2}), F_2 = Q(i, \sqrt[4]{2}), F_3 = Q(i, \sqrt[6]{2}), \ldots$.
Alternate solution. Let $p_1, p_2, p_3, \ldots$ be distinct primes and $F_0 = Q(i), F_1 = Q(i, \sqrt{p_1}), F_2 = Q(i, \sqrt{p_2}), F_3 = Q(i, \sqrt{p_3}), \ldots$.

53. Because the derivative of $x^{p^n} - x$ is -1, $x^{p^n} - x$ has no multiple zeroes in any extension of Z_p.

54. It follows from Theorem 19.9 that $f(x)$ is reducible over F. By Theorem 19.9, one of a_1 or a_2, is in F. If a_2 is in F, then so is a_1. For if $x - a_2$ is in $F[x]$, when we long divide $f(x)$ by $(x - a_2)^2$, the quotient $x - a_1$ is also in $F[x]$.

55. Let $f(x) = c_n x^n + c_{n-1}x^{n-1} + \cdots + c_1 x + c_0$. Then, using Exercise 49 in Chapter 13 and that the fact that $0 = 0^2$ and $1 = 1^2$, we have
$f(a^2) = c_n(a^2)^n + c_{n-1}(a^2)^{n-1} + \cdots + c_1(a^2) + c_0 = (c_n a^n)^2 + (c_{n-1}a^{n-1})^2 + \cdots + (c_1 a)^2 + c_0^2 = (c_n a^n + c_{n-1}a^{n-1} + \cdots + c_1 a + c_0)^2 = f(a)^2 = 0$. By repeating the first part of this problem, we know that the five zeros are a, a^2, a^4, a^8, a^{16}. If any two of these were equal, then we would have $a^k = 1$ for some positive integer k less than 31, whereas $|a| = 31$ because $|(Z_2[x]\langle f(x)\rangle)^*| = 31$.

56. It is a zero of $f(x)$ of multiplicity at least 3.

CHAPTER 20

Algebraic Extensions

1. It follows from Theorem 20.1 that if $p(x)$ and $q(x)$ are both monic irreducible polynomials in $F[x]$ with $p(a) = q(a) = 0$, then $\deg p(x) = \deg q(x)$. If $p(x) \neq q(x)$, then $(p - q)(a) = p(a) - q(a) = 0$ and deg $(p(x) - q(x)) < \deg p(x)$, contradicting Theorem 20.1.

   To prove Theorem 20.3 we use the Division Algorithm (Theorem 16.2) to write $f(x) = p(x)q(x) + r(x)$, where $r(x) = 0$ or $\deg r(x) < \deg p(x)$. Since $0 = f(a) = p(a)q(a) + r(a) = r(a)$ and $p(x)$ is a polynomial of minimum degree for which a is a zero, we may conclude that $r(x) = 0$.

2. If $f(x) \in F[x]$ does not split in E, then it has a nonlinear factor $q(x)$ which is irreducible over E. But then $E[x]/\langle q(x)\rangle$ is a proper algebraic extension of E.

3. Let $F = Q(\sqrt{2}, \sqrt[3]{2}, \sqrt[4]{2}, \ldots)$. Since $[F : Q] \geq [Q(\sqrt[n]{2}) : Q] = n$ for all n, $[F : Q]$ is infinite. To prove that F is an algebraic extension of Q, let $a \in F$. There is some k such that $a \in Q(\sqrt{2}, \sqrt[3]{2}, \sqrt[4]{2}, \ldots, \sqrt[k]{2})$. It follows from Theorem 20.5 that $[Q(\sqrt{2}, \sqrt[3]{2}, \sqrt[4]{2}, \ldots, \sqrt[k]{2}) : Q]$ is finite and from Theorem 20.4 that $Q(\sqrt{2}, \sqrt[3]{2}, \sqrt[4]{2}, \ldots, \sqrt[k]{2})$ is algebraic.

4. Suppose $E(a)$ is an algebraic extension of E. Then a is algebraic over F (Theorem 20.7). So if $f(x)$ is the minimal polynomial for a over F then $f(x)$ can be written in the for $(x - a_1)(x - a_2) \cdots (x - a_n)$ where each $a_i \in E$. But then $f(a) = 0$ implies $a = a_i$ for some i and $a \in E$.

5. Since every irreducible polynomial in $F[x]$ is linear, every irreducible polynomial in $F[x]$ splits in F. So, by Exercise 4, F is algebraically closed.

6. Suppose $g(x) = h(x)k(x)$ where $h(x)$ is irreducible over $F(a)$ and $\deg h(x) < \deg g(x)$. Let b be a zero of $h(x)$ in some extension of $F(a)$. Then $[F(a, b) : F(a)] < \deg g(x) = [F(b) : F]$. Since $[F(a, b) : F] = [F(a, b) : F(a)][F(a) : F] = [F(a, b) : F(b)][F(b) : F]$ we know that $[F(a, b) : F]$ is divisible by $[F(a) : F][F(b) : F]$. But $[F(a, b) : F] = [F(a, b) : F(a)][F(b) : F] < [F(b) : F][F(a) : F]$. This is a contradiction.

7. Suppose $Q(\sqrt{a}) = Q(\sqrt{b})$. If $\sqrt{b} \in Q$, then $\sqrt{a} \in Q$ and we may take $c = \sqrt{a}/\sqrt{b}$. If $\sqrt{b} \notin Q$, then $\sqrt{a} \notin Q$. Write $\sqrt{a} = r + s\sqrt{b}$ where r and s belong to Q. Then $r = 0$ for, if not, then $a = r^2 + 2rs\sqrt{b} + b$ and therefore $(a - r^2 - b)/2r = s\sqrt{b}$. But $(a - r^2 - b)/2r$ is rational whereas $s\sqrt{b}$ is irrational. Conversely, if there is an element $c \in Q$ such that $a = bc^2$ (we may assume that c is positive) then, by Exercise 7 in Chapter 19, $Q(\sqrt{a}) = Q(\sqrt{bc^2}) = Q(c\sqrt{b}) = Q(\sqrt{b})$.

8. Since $(\sqrt{3} + \sqrt{5})^2 \in Q(\sqrt{15})$, $[Q(\sqrt{3} + \sqrt{5}) : Q(\sqrt{15})] = 2$. A basis is $\{1, \sqrt{3} + \sqrt{5}\}$. For the second question, note that because $2^{1/12} = 2^{4/12} \cdot 2^{-3/12} = 2^{1/3} \cdot 2^{-1/4}$, we have that $Q(\sqrt{2}, \sqrt[3]{2}, \sqrt[4]{2}) = Q(\sqrt[12]{2})$. Since $\sqrt[12]{2}$ is a zero of $x^{12} - 2$ and $x^{12} - 2$ is irreducible over Q by Eisenstein, we know that $Q(\sqrt[12]{2})$ has degree 12 over Q and a basis is $1, 2^{1/12}, 2^{2/12}, \ldots, 2^{11/12}$.

Alternate solution. First observe that $Q(\sqrt{2}, \sqrt[3]{2}, \sqrt[4]{2}) = Q(\sqrt[3]{2}, \sqrt[4]{2})$. Because $[Q(\sqrt[3]{2}, \sqrt[4]{2}) : Q] = [Q(\sqrt[3]{2}, \sqrt[4]{2}) : Q(\sqrt[4]{2})][Q(\sqrt[4]{2}) : Q]$ and a basis for $Q(\sqrt[3]{2}, \sqrt[4]{2})$ over $Q(\sqrt[4]{2})$ is $X = \{1, 2^{1/3}, 2^{2/3}\}$ and a basis for $Q(\sqrt[4]{2})$ over Q is $Y = \{1, 2^{1/4}, 2^{2/4}, 2^{3/4}\}$, we know that XY is a basis for $Q(\sqrt[3]{2}, \sqrt[4]{2})$ over Q.

9. Since $[F(a) : F] = 5$, $\{1, a, a^2, a^3, a^4\}$ is a basis for $F(a)$ over F. Also, from $5 = [F(a) : F] = [F(a) : F(a^3)][F(a^3) : F]$ we know that $[F(a^3) : F] = 1$ or 5. However, $[F(a^3) : F] = 1$ implies that $a^3 \in F$ and therefore the elements $1, a, a^2, a^3, a^4$ are not linearly independent over F because a^3 can be written as $1 \cdot a^3$ and as $a^3 \cdot 1$. So, $[F(a^3) : F] = 5$.

10. Since $[E : F] = [E : F(a)][F(a) : F]$ we have $[F(a) : F] = [E : F]$, in which case $F(a) = E$, or $[F(a) : F] = 1$, in which case $F(a) = F$.

11. If a is a zero of $f(x)$ in E then $n = [E : F] = [E : F(a)][F(a) : F] = [E : F](\deg f(x))$.

12. Observe that $[F(a) : F(a^2)] = 1$ or 2 and that $n = [E : F] = [E : F(a)][F(a) : F(a^2)][F(a^2) : F]$.

13. $g(x) = f(x/b - c/b)$

14. Since β is a zero of $x^n - \beta^n$, $F(\beta)$ is an algebraic extension of $F(\beta^n)$, and $1, \beta, \beta^2, \ldots, \beta^{n-1}$ is a basis for $F(\beta)$ over $F(\beta^n)$, we have $[F(\beta) : F(\beta^n)] = n$. For $n = 4$, we have $3\beta^5 + 2\beta^{-3} = (3\beta^2)\beta^3 + 2\beta^{-4}\beta$.

15. By the Primitive Element Theorem there is an element b in $F(a_1, a_2)$ such that $F(a_1, a_2) = F(b)$. Then, by induction on n, there is an element c in $F(b, a_3, \ldots, a_n)$ such that $F(c) = F(b, a_3, \ldots, a_n) = F(a_1, a_2, \ldots, a_n)$.

16. Because $11\sqrt{12} - 7\sqrt{45} = 22\sqrt{3} - 21\sqrt{5}$ belongs to $Q(\sqrt{3}, \sqrt{5})$ and $[Q(\sqrt{3} + \sqrt{5} : Q] = [Q(\sqrt{3}, \sqrt{5}) : Q] = 4$ we know that $\{1, \sqrt{3} + \sqrt{5}, (\sqrt{3} + \sqrt{5})^2, (\sqrt{3} + \sqrt{5})^3\}$ is a basis for $Q(\sqrt{3}, \sqrt{5})$.

17. 6; 3; 2.

18. $1, 2^{1/6}, 2^{2/6}, 2^{3/6}, 2^{4/6}, 2^{5/6}$; $1, 2^{1/6}, 2^{2/6}$; $1, 2^{1/6}$.

19. They are lattice isomorphic.

20. If $b = 0$, then $x - c$ is minimal. If $b \neq 0$ for $g(x) = f((x - c)/b)$ we have $g(ab + c) = f((ab + c - c)/b) = f(a) = 0$. That $g(x)$ has minimum degree follows from the fact that $F(a) = F(ab + c)$ (see Exercise 7 Chapter 19). Or that $g(bx + c) = f(x)$.

21. Elements of $Q(\pi)$ have the form $(a_m\pi^m + a_{m-1}\pi^{m-1} + \cdots + a_0)/(b_n\pi^n + b_{n-1}\phi^{n-1} + \cdots + b_0)$, where the a's and b's are rational numbers. So, if $\sqrt{2} \in Q(\pi)$, we have an expression of the form $2(b_n\pi^n + b_{n-1}\phi^{n-1} + \cdots + b_0)^2 = (a_m\pi^m + a_{m-1}\pi^{m-1} + \cdots + a_0)^2$. Equating the lead terms of both sides, we have $2b_n^2\pi^{2n} = a_m^2\pi^{2m}$. But then we have $m = n$, and $\sqrt{2}$ is equal to the rational number a_m/b_n. This argument works for any real number of the form $\sqrt[k]{a}$ not in Q.

22. A splitting field of a polynomial in $Q[x]$ is an algebraic extension of Q whereas **C** is a transcendental extension of Q.

23. Let $\{x_1, x_2, \ldots, x_n\}$ be a basis for K over F. Then $K = F(x_1, x_2, \ldots, x_n)$ for some $x_1, x_2, \ldots, x_n$ in K.

24. If an irreducible polynomial $p(x)$ in $\mathbf{R}[x]$ has degree n and a is a zero in $\mathbf{C}$ of $p(x)$ then $2 = [\mathbf{C}:\mathbf{R}] = [\mathbf{C}: \mathbf{R}(a)][\mathbf{R}(a):\mathbf{R}] = [\mathbf{C}: \mathbf{R}(a)]n$. So, $n = 1$ or 2.

25. By Theorem 20.5, we have that $[F_3 : F_1] = [F_3 : F_2][F_2 : F_1]$ and $[F_4 : F_1] = [F_4 : F_3][F_3 : F_1] = [F_4 : F_3][F_3 : F_2][F_2 : F_1]$. The general case that $[F_n : F_1] = [F_n : F_{n-1}][F_{n-1}] : F_{n-2}] \cdots [F_2 : F_1]$ follows by induction.

26. Suppose that $[E : F] = 1$. Because $\{1\}$ is a linearly independent set over F, it is a basis for E over F. So every element of E has the form $a \cdot 1 = a$ for some a in F. Now suppose that $E = F$. Then $\{1\}$ is a basis for E over F.

27. Since $[L : F] = [L : K][K : F]$ we have $[K : F] = 1$. It follows that $F = K$.

28. Pick a in K but not in F. Now use Theorem 20.5.

29. If a is a zero of $p(x) \in Q[x]$, then for $k \geq 2$, $\sqrt[k]{a}$ is a zero of $p(x^k) \in Q[x]$.
Alternate solution. Observe that
$[Q(\sqrt[k]{a} : Q] = [Q(\sqrt[k]{a} : Q(a)][Q(a) : Q] \leq k[Q(a) : Q]$ is finite. For the case of $a^{m/n}$ where m and n are positive integers, note that a is algebraic over Q implies that a^m is algebraic over Q. So $(\sqrt[n]{(a^m)} = a^{m/n}$ is algebraic over Q.

30. Note that if $c \in Q(\beta)$ and $c \notin Q$, then $5 = [Q(\beta) : Q] = [Q(\beta) : Q(c)][Q(c) : Q]$ so that $[Q(c) : Q] = 5$. On the other hand, $[Q(\sqrt{2}) : Q] = 2$, $[Q(\sqrt[3]{2}) : Q] = 3$, and $[Q(\sqrt[4]{2}) : Q] = 4$.

31. Since $\sqrt{2} = (\sqrt[6]{2})^3$ and $\sqrt[3]{2} = (\sqrt[6]{2})^2$ we know that $Q(\sqrt{2}, \sqrt[3]{2}) \subseteq Q(\sqrt[6]{2})$. Because $2^{1/6} = 2^{1/2}(2^{1/3})^{-1}$, we also have $Q(\sqrt[6]{2}) \subseteq Q(\sqrt{2}, \sqrt[3]{2})$.
Alternate solution. Since $\sqrt{2} = (\sqrt[6]{2})^3$ and $\sqrt[3]{2} = (\sqrt[6]{2})^2$ we know that $Q(\sqrt{2}, \sqrt[3]{2}) \subseteq Q(\sqrt[6]{2})$. Also, $[Q(\sqrt{2}, \sqrt[3]{2}) : Q]$ is divisible by $[Q(\sqrt{2}) : Q] = 2$ and $[Q(\sqrt[3]{2}) : Q] = 3$. It follows that $[Q(\sqrt{2}, \sqrt[3]{2}) : Q] = 6 = [Q(\sqrt[6]{2}) : Q]$.

32. By closure, $Q(\sqrt{a} + \sqrt{b}) \subseteq Q(\sqrt{a}, \sqrt{b})$. Since $(\sqrt{a} + \sqrt{b})^{-1} = \frac{1}{\sqrt{a}+\sqrt{b}} \frac{\sqrt{a}-\sqrt{b}}{\sqrt{a}-\sqrt{b}} = \frac{\sqrt{a}-\sqrt{b}}{a-b}$ and $a - b \in Q(\sqrt{a} + \sqrt{b})$ we have $\sqrt{a} - \sqrt{b} \in Q(\sqrt{a} + \sqrt{b})$. (The case that $a - b = 0$ is trivial.) It follows that $\sqrt{a} = \frac{1}{2}((\sqrt{a} + \sqrt{b}) + (\sqrt{a} - \sqrt{b}))$ and $\sqrt{b} = \frac{1}{2}((\sqrt{a} + \sqrt{b}) - (\sqrt{a} - \sqrt{b}))$ are in $Q(\sqrt{a}, \sqrt{b})$. So, $Q(\sqrt{a}, \sqrt{b}) \subseteq Q(\sqrt{a} + \sqrt{b})$.

33. 9

34. Suppose $E_1 \cap E_2 \neq F$. Then $[E_1 : E_1 \cap E_2][E_1 \cap E_2 : F] = [E_1 : F]$ implies $[E_1 : E_1 \cap E_2] = 1$, so that $E_1 = E_1 \cap E_2$. Similarly, $E_2 = E_1 \cap E_2$.

35. First observe that $\sqrt[4]{3}$ is a zero of $x^2 - \sqrt{3}$, which is in $Q(\sqrt{3})$. If this polynomial is not a minimal polynomial of $\sqrt[4]{3}$ over $Q(\sqrt{3})$, then $[Q(\sqrt[4]{3}) : Q(\sqrt{3})] = 1$. But then, $4 = [Q(\sqrt[4]{3}) : Q] = [Q(\sqrt[4]{3}) : Q(\sqrt{3})][Q(\sqrt{3}) : Q] = 1 \cdot 2$, which is a contradiction.

36. Observe that $F(1 + a^{-1}) = F(a^{-1}) = F(a)$.

37. If ab is a zero of $c_n x^n + \cdots + c_1 x + c_0 \in F[x]$, then a is a zero of $c_n b^n x^n + \cdots + c_1 bx + c_0 \in F(b)[x]$.
Alternate proof. Observe that because both ab and b are algebraic over $F(b)$, so is $a = (ab)b^{-1}$ (Corollary of Theorem 20.7.)

38. It suffices to show that for every non-zero element a in R a^{-1} is also in R. Since a is in E, it is the zero of some minimal polynomial in $F[x]$ of degree d. By closure under multiplication, we know that the basis $\{1, a, a^2, \ldots, a^{d-1}\}$ of $F(a)$ is contained in R. So, $F(a) \subseteq R$ and $a^{-1} \in F(a)$.

39. Note that $x-3, x^3-(\pi^3)^2$ and $x-\pi^3$ are in $Q(\pi^3)[x]$ and have $3, \pi^2$, and π as zeros, respectively. So, $3, \pi^2$, and π are algebraic over $Q(\pi^3)$. Then, we have by the corollary to Theorem 20.7 that $3\pi^2+\pi$ is algebraic over $Q(\pi^3)$.

40. Every element of $F(a)$ can be written in the form $f(a)/g(a)$, where $f(x), g(x) \in F[x]$. If $f(a)/g(a)$ is algebraic and not a member of F, then there is some $h(x) \in F[x]$ such that $h(f(a)/g(a)) = 0$. By clearing fractions and collecting like powers of a, we obtain a polynomial in a with coefficients from F equal to 0. But then a would be algebraic over F.

41. Note that $[F(a,b):F]$ is divisible by both $m = [F(a):F]$ and $n = [F(b):F]$ and $[F(a,b):F] = [F(a,b):F(b)][F(b):F] \leq [F(a):F][F(b):F] \leq mn$. So, $[F(a,b):F] = mn$.

42. Since a is a zero of x^3-a^3 over $F(a^3)$, we have $[F(a):F(a^3)] \leq 3$. For the second part, take $F=Q, a=1$; $F=Q, a=(-1+i\sqrt{3})/2; F=Q, a=\sqrt[3]{2}$.

43. Take $F=Q$, $a=\sqrt[4]{2}$, $b=\sqrt[6]{2}$. Then $[F(a,b):F]=12$ and $[F(a):F][F(b):F]=24$.

44. Since E must be an algebraic extension of $\mathbf{R}$, we have $E \subseteq \mathbf{C}$ and so $[\mathbf{C}:E][E:\mathbf{R}] = [\mathbf{C}:\mathbf{R}] = 2$. If $[\mathbf{C}:E]=2$, then $[E:\mathbf{R}]=1$ and therefore $E=\mathbf{R}$. If $[\mathbf{C}:E]=1$, then $E=\mathbf{C}$.

45. Use the quadratic formula and Exercise 7 of Chapter 19.

46. Let a be a zero of $p(x)$ in some extension of F. First note $[E(a):E] \leq [F(a):F] = \deg p(x)$. Then observe that $[E(a):F(a)][F(a):F] = [E(a):F] = [E(a):E][E:F]$. This implies that $\deg p(x)$ divides $[E(a):E]$, so that $\deg p(x) = [E(a):E]$. It now follows from Theorem 19.3 that $p(x)$ is irreducible over E.

47. If E is a finite extension of F, pick $a_1 \in E$ but not in F. Then $[E:F] \geq [F(a_1):F]$. If $F(a_1) \neq E$, pick $a_2 \in E$ not in $F(a_1)$. Then $[E:F] \geq [F(a_1,a_2):F]$. Since $[E:F]$ is finite, we can continue this process only a finite number of times. The converse follows from Theorem 20.5.

48. Suppose that $\alpha+\beta$ and $\alpha\beta$ are algebraic over Q and that $\alpha \geq \beta$. Then $\sqrt{(\alpha+\beta)^2-4\alpha\beta} = \sqrt{\alpha^2-2\alpha\beta+\beta^2} = \sqrt{(\alpha-\beta)^2} = \alpha-\beta$ is also algebraic over Q. Also, $\alpha = ((\alpha+\beta)-(\alpha-\beta))/2$ is algebraic over Q, which is a contradiction.

49. If $f(a)$ is a zero of $g(x) \in F[x]$, then a is a zero of $(g \circ f)(x) \in F[x]$.

50. It follows from the Quadratic Formula that $\sqrt{b^2-4ac}$ is a primitive element.

51. Let $\deg f(x) = m$, $\deg g(x) = n$ and suppose that $f(x)$ is irreducible over $F(b)$. Then $m = [F(a,b):F(b)]$ and $[F(a,b):F] = [F(a,b):F(b)][F(b):F] = mn = [F(a,b):F(a)][F(a):F] = [F(a,b):F(a)]m$. Thus, $n = \deg g(x) = [F(a,b):F(a)]$ and therefore $g(x)$ is irreducible over $F(a)$. The other half follows by symmetry.

52. Because $a \in Q(\sqrt{a})$ it suffices to show that $\sqrt{a} \in Q(a)$. Since $a^3=1$ we have $a^4=a$. Then $a^2 = \sqrt{a} \in Q(a)$.

53. By the Factor Theorem (Corollary 2 of Theorem 16.2), we have $f(x) = (x-a)(bx+c)$, where $a, b, c \in F(a)$.

54. Say a is a generator of F^*. F cannot have characteristic 0 because the subgroup of rationals is not cyclic. Thus $F = Z_p(a)$, and by Theorem 20.3 it suffices to show that a is algebraic over Z_p. If $a \in Z_p$, we are done. Otherwise, $1 + a = a^k$ for some $k \neq 0$. If $k > 0$, we are done. If $k < 0$, then $a^{-k} + a^{1-k} = 1$ and we are done.

55. Suppose that there are such rationals. Let $s = s_1/s_2$ and $t = t_1/t_2$ where s_1, s_2, t_1, t_2 are positive integers. Then $(r^s)^{s_2t_2} = (\pi^t)^{s_2t_2}$. But the left side is rational and the right side is not.

56. If $[K : F] = n$, then there are elements $v_1, v_2, \ldots, v_n$ in K that constitute a basis for K over F. The mapping $a_1v_1 + \cdots + a_nv_n \to (a_1, \ldots, a_n)$ is a vector space isomorphism from K to F^n. If K is isomorphic to F^n, then the n elements in K corresponding to $(1, 0, \ldots, 0)$, $(0, 1, \ldots, 0)$, $\ldots$, $(0, 0, \ldots, 1)$ in F^n constitute a basis for K over F.

57. A counterexample is $[Q(4^{1/4}) : Q] = 2$; another is $[Q(4^{1/6} : Q] = 3$.

58. Observe that $[F(a, b) : F(a)] \leq [F(a, b) : F(a)][F(a) : F] = [F(a, b) : F]$.

59. First note that $c_0 \neq 0$, for otherwise $c_{d-1}x^{d-1} + c_{d-2}x^{d-2} + \cdots + c_1x + c_0$ has x as a factor over field F. So from $c_{d-1}a^{d-1} + c_{d-2}a^{d-2} + \cdots + c_1a + c_0 = 0$ we have $-c_0 = a(c_{d-1}a^{d-2} + c_{d-2}a^{d-3} + \cdots + c_1)$. Multiplying both sides by $a^{-1}(-c_0)^{-1}$ we obtain $a^{-1} = -c_0^{-1}c_{d-1}a^{d-2} - c_0^{-1}c_{d-2}a^{d-3} - \cdots - c_0^{-1}c_1$.

60. Observe that $K = F(a_1, a_2, \ldots, a_n)$, where $a_1, a_2, \ldots, a_n$ are the zeros of the polynomial. Now use Theorem 20.5.

CHAPTER 21

Finite Fields

1. Since $729 = 9^3$, $[GF(729) : GF(9)] = 3$; since $64 = 8^2$, $[GF(64) : GF(8)] = 2$.

2. Use Theorem 20.5.

3. The lattice of subfields of GF(64) looks like Figure 20.3 with GF(2) at the bottom, GF(64) at the top, and GF(4) and GF(8) on the sides.

4. First observe that $0 = a^3 + a^2 + 1 = a^2(a+1) + 1$ so $(a+1)^{-1} = a^2$. Then solving the equation for x we have that $x = a^2 + a$.

5. $GF(2^6)$

6. For any a in $GF(32)^*$ $|a|$ must divide 31.

7. By Theorem 21.3, the other solution is $a^3 = 2a + 1$.

8. Since a is a zero and the prime subfield has characteristic 2, we have from Theorem 21.3 that the other zeros are a^2 and a^4.

9. Since each binomial coefficient $\binom{p^i}{j}$ other than $j = 1$ and $j = p^i$ is divisible by p, we have $(a+b)^{p^i} = a^{p^i} + b^{p^i}$. Clearly $(ab)^{p^i} = a^{p^i} b^{p^i}$. Since $GF(p^{p^i})$ is a field, $a^{p^i} = 0$ only when $a = 0$, so the Ker $\phi = \{0\}$.

10. It follows from the Finite Subgroup Test (Theorem 3.3)) that S is a group under addition and the non-zero elements form a group under multiplication.

11. Observe that $x^2 + 1 = (x+1)^2$.
Alternate solution. Any solution a has the property that $a^2 = -1 = 1$. But because $|GF(2^n)^*| = 2^n - 1$, there is no element in $GF(2^n)^*$ of order 2.

12. Note the $|a^5| = 3$.

13. Let a be a zero of a cubic irreducible $f(x)$ in $Z_2[x]$ in some extension of Z_2. Then $Z_2(a)$ has order 8, and by Theorem 21.3, the other two zeros of $f(x)$ are a^2 and a^4, which also belong to $Z_p(a)$. So, $f(x)$ splits over $Z_p(a)$.

14. Use Theorem 21.1.

15. Because $GF(64)^*$ is a cyclic group of order 63, we have by Theorem 4.4 that there are $\phi(63) = \phi(9)\phi(7) = 36$ elements with the property.

16. According to Theorem 21.3, the zeros are: a, a^2, a^4, a^8 for the first part and $a^7, (a^7)^2, (a^7)^4, (a^7)^8 = a^7, a^{14}, a^{13}, a^{11}$ for the second part.

17. Let $|F| = p^n$. Then n must be divisible by both 4 and 6. So, by Theorem 21.4, F must also have subfields of order p^{12}, p^3, p^2 and p.

18. By Theorem 22.3, m divides 3.

19. By observation, the linear factors are x and $x-1$. Since $x^8 - x$ has no multiple zeros in GF(8), it has no multiple irreducible factors over GF(2). Because the degrees of the irreducible factors must divide 3 and sum to 8 it follows that the nonlinear irreducible factors must be two cubics. Noting that the only cubic irreducibles over Z_2 are $x^3 + x + 1$ and $x^3 + x^2 + 1$, these are the remaining factors of $x^8 - x$.

20. Note that by Theorem 21.4, the only proper subfield of GF(32) is GF(2). So, the degree of a over GF(2) is 5.

21. Observe that $a^{16} = a^4$ implies that $0 = a^{16} - a^4 = (a^4)^4 - a^4 = (a^4 - a)^4$.

22. Observe that $Z_p(b)$ is a subfield field of $Z_p(a)$ and $|Z_p(b)| = p^n$.

23. First note that $x, x-1$ and $x-2$ are factors and they are the only linear factors. If $p(x)$ is an irreducible factor of $x^{27} - x$ and a is a zero of an irreducible factor $x^{27} - x$ of degree d, then $Z_3(a)$ is a subfield of GF(27) of order 3^d, and by Theorem 21.4 we have that $d = 1$ or 3. So, the irreducible factorization of $x^{27} - x$ consists of three linear factors and eight cubic factors.

24. Observe that the mapping from F to itself that takes x to x^2 is a field automorphism with kernel $\{0\}$. Since a 1-1 mapping from a finite set to itself is onto we are done.

25. By Theorem 21.4, GF(p^n) is properly contained in GF(p^m) when n is a proper divisor of m. So the smallest such field is GF(p^{2n}).

26. Observe that every element of the GF(p^n) is a zero of $x^{p^n} - x$. So evaluating $x^{p^n} - x + 1$ for any element of GF(p^n) gives 1.

27. To show that F is a field, let $a, b \in F$. Then $a \in F_i$ for some i and $b \in F_j$ for some j and $a, b \in F_k$ where k is the maximum of i and j. It follows that $a - b \in F_k, ab \in F_k$ and $a^{-1} \in F_k$ when $a \neq 0$.

28. GF(2^{10}), GF(2^{15}), GF(2^{25}).

29. Since 0, 1, and 2 are zeros, we know $x, x-1$ and $x-2$ are factors and by Theorem 19.8 these each have multiplicity 1. Theorem 21.5 tells us that all other irreducible factors have degree 2 and, since their degrees must sum to six, there are three of them. Finally, from the proof of Theorem 21.1 we know that each of the nine elements of GF(9) are zeros of $x^9 - x$. So no two quadratic reducibles can be the same. The factorization $x^9 - x = x(x-1)(x-2)(x^2+1)(x^2+x+2)(x^2+2x+2)$.

30. It follows from Theorem 21.3 that the desired field is GF(2^{12}).

31. Since $|(Z_3[x]/\langle x^3 + 2x + 1\rangle)^*| = 26$, we need only show that $|x| \neq 1, 2$ or 13. Obviously, $x \neq 1$ and $x^2 \neq 1$. Using the fact that $x^3 + 2x + 1 = 0$ and doing the calculations we obtain $x^{13} = 2$.

32. The argument in the proof of Theorem 21.3 shows that 4 divides $5^n - 1$. Since GF(5^n)* is a cyclic group of order $5^n - 1$ and $|2| = 4$, we know that $\langle 2\rangle$ is the unique subgroup of order 4. We also know that $a^{(5^n-1)/4}$ is an element of order 4. By Corollary 3 of Theorem 4.3, if b is an element of a finite cyclic group of order 4 then the only other element in the group of order 4 is b^3. So, $k = (5^n - 1)/4$.

33. From $x^3 = x + 1$ we get $x^9 = (x+1)^3 = x^3 + 1 = x + 2$ and $x^4 = x^2 + x$. So, $x^{13} = (x + 2(x^2 + 2) = 1$. Then $|2x| = 26$ and is a generator.

34. Since $1, x$ and x^2 form a basis for $Z_3[x]/\langle f(x)\rangle$ over Z_3 we have that $|x| \neq 2$; for otherwise $x^2 - 1 = (x-1)(x+1) = 0$ and therefore $x = \pm 1$. By Lagrange's Theorem $|x|$ is 13 or 26. If $|x| = 13$, then $|2x| = 26$.

35. Note that if K is any subfield of $\mathrm{GF}(p^n)$, then K^* is a subgroup of the cyclic group $\mathrm{GF}(p^n)^*$. So, by Theorem 4.3, K^* is the unique subgroup of $\mathrm{GF}(p^n)^*$ of its order.

36. Observe that $|ab| = |\langle a\rangle\langle b\rangle| = |\langle a\rangle||\langle b\rangle|/|\langle a\rangle \cap \langle b\rangle| = |\langle a\rangle||\langle b\rangle| = 5 \cdot 16 = 80$.

37. Let $a, b \in K$. Then, by Exercise 49b in Chapter 13, $(a-b)^{p^m} = a^{p^m} - b^{p^m} = a - b$. Also, $(ab)^{p^m} = a^{p^m} b^{p^m} = ab$. So, K is a subfield.

38. Mimic the proof of the case where the field is $\mathrm{GF}(p^n)$ and the group is $\mathrm{GF}(p^n)^*$ given in Theorem 21.2.

39. By Corollary 4 of Lagrange's Theorem (Theorem 7.1), for every element a in F^* we have $a^{p^n-1} = 1$. So, every element in F^* is a zero of $x^{p^n} - x$.

40. Theorem 21.3 reduces the problem to constructing the subgroup lattices for Z_{18} and Z_{30}.

41. They are identical to the lattice of Z_{30}.

42. Without loss of generality, we may assume that $p(x)$ is monic. In some splitting field K of $p(x)$ we can write $p(x) = (x-a_1)(x-a_2)\cdots(x-a_m)$ where the a_i are distinct and $K = \mathrm{GF}(p)(a_1, a_2, \ldots, a_m)$. Then $K = GF(p^n)$ for some n and since every element of K is a zero of $x^{p^n} - x$ we have that $p(x) = (x-a_1)(x-a_2)\cdots(x-a_m)$ divides $x^{p^n} - x$.

43. The hypothesis implies that $g(x) = x^2 - a$ is irreducible over $\mathrm{GF}(p)$. Then a is a square in $\mathrm{GF}(p^n)$ if and only if $g(x)$ has a zero in $\mathrm{GF}(p^n)$. Since $g(x)$ splits in $\mathrm{GF}(p)[x]/\langle g(x)\rangle \approx \mathrm{GF}(p^2)$, $g(x)$ has a zero in $\mathrm{GF}(p^n)$ if and only if $\mathrm{GF}(p^2)$ is a subfield of $\mathrm{GF}(p^n)$. The statement now follows from Theorem 21.3.

44. Let a be a zero of $f(x)$ in $E = Z_p[x]/\langle f(x)\rangle$. Then $|Z_p(a)| = p^3$ and, by the corollary of Theorem 21.3, $Z_p(a)$ is the splitting field of $f(x)$.

45. F^*; $\mathrm{GF}(2^5)^* = \langle a^{33}\rangle$; $\mathrm{GF}(2^2)^* = \langle a^{341}\rangle$; $\mathrm{GF}(2)^* = \langle a^{1023}\rangle = \{1\}$.

46. Note that $\mathrm{GF}(p^{2n})$ is an algebraic extension of $\mathrm{GF}(p^n)$.
Alternate proof. Let $F = \{a_1, a_2, \ldots, a_n\}$ be a finite field. Then $f(x) = (x-a_1)(x-a_2)\cdots(x-a_n) + 1 \in F[x]$ but has no zeros in F.

47. Since both a^{62} and -1 have order 2 in the cyclic group F^* and a cyclic group of even order has a unique element of order 2 (see Theorem 4.4), we have $a^{62} = -1$.

48. p^k where $k = \gcd(s, t)$.

49. If K is a finite extension of a finite field F, then K itself is a finite field. So, $K^* = \langle a\rangle$ for some $a \in K$ and therefore $K = F(a)$.

50. Suppose b is one solution of $x^n = a$. Since F^* is a cyclic group of order $q-1$, it has a cyclic subgroup of order n, say $\langle c\rangle$. Then each member of $\langle c\rangle$ is a solution to the equation $x^n = 1$. It follows that $b\langle c\rangle$ is the solution set of $x^n = a$.

51. Observe that $p - 1 = -1$ has multiplicative order 2 and $a^{(p^n-1)/2}$ is the unique element in $\langle a\rangle$ of order 2.

52. Since d divides n, GF(p^n) contains the subfield GF(p^d), and by Corollary 2 of Theorem 21.2, GF(p^d) has the form GF$(p)(a)$ for some element a in GF(p^d) of degree d. Moreover, by Theorem 21.1, GF$(p)(a)$ has the form GF$(p)[x]/\langle p(x)\rangle$ where $p(x)$ is an irreducible polynomial over GF(p), the degree of $p(x)$ is d, and $p(a) = 0$. Since $p(x)$ is irreducible over GF(p) the only common divisor of $p(x)$ and $x^{p^n} - x$ is 1 or $p(x)$. If the common divisor is 1 then by Exercise 43 of Chapter 16 there are elements $h(x)$ and $k(x)$ of GF$(p)[x]$ such that $p(x)h(x) + (x^{p^n} - x)k(x) = 1$. But since $x - a$ divides the left side we have a contradiction.

53. Observe that the mapping from the cyclic group GF$(p^n)^*$ to itself that takes x to x^2 is a group homomorphism with the kernel $\{\pm 1\}$. So, the mapping is 2-1.

54. Since 5 mod 4 = 1, we have that $5^n - 1$ is divisible by 4 for all n. Now observe that 2 has multiplicative order 4 and $a^{(5^n-1)/4}$ has order 4. (The only other element of order 4 is $a^{3(5^n-1)/4}$.)

55. Since $p = 1 \pmod 4$ we have $p^n = 1$ mod 4 and GF$(p^n)^*$ is a cyclic of order $p^n - 1$. So, by Theorem 4.4 there are exactly two elements of order 4.

56. When n is odd, $p^n = 3$ mod 4 and therefore $p^n - 1$ is not divisible by 4. By Theorem 4.3 $\langle a\rangle$ has no element of order 4. When $n = 2m$, $p^n = (p^2)^m = (3^2)^m = 1$ mod 4 and therefore $p^n - 1$ is divisible by 4. Thus, by Theorem 4.4, GF$(p^n)^*$ has exactly two elements of order 4.

57. First note that a is not in Z_5, for then $x - a$ would be a factor of the irreducible. Then taking $b = 0$ and we solve for $c = (4a + 1)(3a + 2)^{-1}$.

58. Since $|(Z_2[x]/\langle x^4 + x^2 + 1\rangle)^*| = 15$ we need only check to see that a^3 and a^5 are not 1. $a^3 \neq 1$ because a has degree 4. $a^4 = a^2 + 1$ implies that $a^5 = a^3 + a \neq 1$ because $1, a, a^2$, and a^3 are basis elements.

59. It is a field of order 4^5.

60. Because 2 = [GF(64):GF(8)] = [GF(64):GF(8)(a)][GF(8)(a)][GF(8)] we have deg $f(x) = 2$. Let GF$(64)^* = \langle a\rangle$. Then GF$(8)^* = \langle a^9\rangle$ and $f(x) = x^2 - a^9 \in$ GF(8) and is irreducible over GF(8) (its zeros are $\pm a^3$).

61. For the case GF(p^n), observe that if $1 + a + a^2 + a^3 + \cdots + a^i = 1 + a + a^2 + a^3 + \cdots + a^j$ for some $i < j$, then $0 = a^{i+1} + \cdots + a^j = a^{i+1}(1 + a + a^2 + \cdots + a^{j-i-1})$ and therefore a^{i+1} is a zero-divisor.

62. By Theorem 21.4 the zeros are $a, a^2, a^{(2^2)}, a^{(2^3)}, a^{(2^4)}$.

63. Let $F_1 = \text{GF}(p^n), F_2 = \text{GF}(p^{2n}), F_3 = \text{GF}(p^{4n}), F_4 = \text{GF}(p^{8n}), \ldots$.

64. Since every 5th degree monic irreducible polynomial over Z_3 splits in GF(3^5), they all divide $x^{3^5} - x$. No such factor can appear more than once in the factorization because $x^{3^5} - x$ has no multiple zeros. By Theorem 21.5, the monic irreducible factors of $x^{3^5} - x$ have degrees 1 or 5 and exactly three have degree 1. So, the remaining irreducible factors have degree 5 and there must be exactly 48 of them in order for the degree of the product of all the irreducible monic factors to be 243.

65. The algebraic closure of Z_2.

66. The only finite subfield of $Z_2(x) = \{f(x)/g(x) \mid f(x), g(x) \in Z_2[x], g(x) \neq 0\}$ (the field of quotients of $Z_2[x]$) is Z_2.

67. By Theorem 21.4, for each prime q the only proper subfield of $\mathrm{GF}(p^q)$ is $\mathrm{GF}(p)$.

68. Since $[F(a,b):F] = [F(a,b):F(a)][F(a):F]$ is finite we know that $F(a,b)$ is finite and therefore $F(a,b)^*$ is a cyclic group. Then, for any generator c of $F(a,b)^*$, we have $F(a,b) = F(c)$.

69. To see that $\mathrm{GF}(p^\infty)$ is a field, let a and b be elements in $\mathrm{GF}(p^\infty)$. Then a belongs to $\mathrm{GF}(p^{m!})$ and b belongs to $\mathrm{GF}(p^{n!})$ for some positive integers m and n. Thus, a and b belong to the larger of these two fields, say $\mathrm{GF}(p^{m!})$, which means that $a-b$ and $ab^{-1} (b \neq 0)$ also belong to $\mathrm{GF}(p^{m!})$, which is a subfield of $\mathrm{GF}(p^\infty)$. Lastly, if $f(x)$ is any polynomial in $\mathrm{GF}(p^\infty)[x]$, then $f(x)$ belongs $\mathrm{GF}(p^{s!}[x]$ for some positive integer s. Thus, the splitting field of $f(x)$ over $\mathrm{GF}(p^{s!})$ is a finite field of characteristic p and so is contained in $\mathrm{GF}(p^t)$ for some positive integer t. This means this splitting field is a subfield of $\mathrm{GF}(p^\infty)$.

70. It follows from the Primitive Element Theorem and Theorem 21.2 that F is an infinite field of prime characteristic.

71. Since $x^{16} - x$ has 16 distinct zeros in its splitting field E, no irreducible factor $p(x)$ of it can occur in the factorization more than once. Moreover, we know from Theorem 16.5 that the possible degrees of $p(x)$ are 1, 2, and 4. Because the factorization as irreducibles includes x and $x-1$, the sum of the degrees of the nonlinear factors is 14. Since 14 is not a multiple of 4, $x^{16} - x$ must have has a quadratic factor. Thus it is $x^2 + x + 1$ because it is the only irreducible quadratic over Z_2. So, that leaves three irreducible fourth- degree factors because the sum of the degrees of the irreducible factors is 16.

72. Suppose there is such an element a in some field extension of $(\mathrm{GF})(p)$. Then both 1 and a are zeros of $x^k - 1 \in F[x]$. So, we can write $x^k - 1 = (x-1)g(x)$. This gives us that $g(a) = 0$, which means deg $a < d$ over F.

73. They are the same.

74. If there is some other element a in $\mathrm{GF}(3^n)$ that is a zero, then $a^3 = -1$ and therefore $|a| = 6$. But 6 does not divide $3^n - 1$, which is the order of the cyclic group $\mathrm{GF}(3^n)^*$.

75. $\mathrm{GF}(5^{3k})$ where k is any positive integer.

76. By Theorem 21.3, we have that $p(x) = x(x-a)(x-a^p)(x-a^{p^2}) \cdots (x-a^{p^d})$.

77. By the proof of Theorem 21.1, we have that $x^n - x = x(x-a)(x-a^2)(x-a^3) \cdots (x-a^{n-1})$.

CHAPTER 22

Geometric Constructions

1. To construct $a + b$, first construct a. Then use a straightedge and compass to extent a to the right by marking off the length of b. To construct $a - b$, use the compass to mark off a length of b from the right end point of a line of length a. The remaining segment has length $a - b$.

2. Let x denote the length of the long side of the triangle. Then $\frac{a}{1} = \frac{x}{b}$.

3. Let y denote the length of the hypotenuse of the right triangle with base 1 and x denote the length of the hypotenuse of the right triangle with the base $|c|$. Then $y^2 = 1 + d^2$, $x^2 + y^2 = (1 + |c|)^2$ and $|c|^2 + d^2 = x^2$. So, $1 + 2|c| + |c|^2 = 1 + d^2 + |c|^2 + d^2$, which simplifies to $|c| = d^2$.

4. Let x denote the length of the side of along the base of the small triangle in figure in the text. Then x is constructible and $\frac{x}{1} = \frac{a}{b}$.

5. Suppose that $\sin\theta$ is constructible. Then, by Exercises 1, 2, and 3, $\sqrt{1 - \sin^2\theta} = \cos\theta$ is constructible. Similarly, if $\cos\theta$ is constructible then so is $\sin\theta$.

6. Look at Figure 22.1.

7. From the identity $\cos 2\theta = 2\cos^2\theta - 1$ we see that $\cos 2\theta$ is constructible if and only if $\cos\theta$ is constructible.

8. By Exercises 5 and 7, it suffices to show that a 30° is constructible. This true because $\cos(30°) = \sqrt{3}/2$.

9. Use Exercise 6.

10. The discussion in this chapter shows that $\cos 20°$ is not constructible. Now use Exercises 5, 6, and 7.

11. Note that solving two linear equations with coefficients in F involves only operations under which F is closed.

12. Say the line is $ax + by + c = 0$ and the circle is $x^2 + y^2 + dx + ey + f = 0$. We seek the simultaneous solution of these two equations. If $a = 0$, then $y = -c/b$ and the equation of the circle reduces to a quadratic in x with coefficients from F. Since the solution of a quadratic involves only the operations of F and a square root of an element from F, the value for x lies in F or in $F(\sqrt{\alpha})$ where $\alpha \in F$ and $\alpha > 0$. If $a \neq 0$, the x terms in the circle can be replaced by $(-b/a)y - c/a$ to obtain a quadratic equation. Then, as in the previous case, y and therefore x lie in F or in $F(\sqrt{\alpha})$.
    For the second portion, consider $x^2 + y^2 = 1$ and $x - y = 0$.

13. This follows from Theorem 17.1 and the mod 5 irreducibility test. (Theorem 17.3.)

14. If the polygon is constructible, so is $\cos(2\pi/7)$. Thus, it suffices to show that $8x^3 + 4x^2 - 4x - 1$ is irreducible over Q. This follows from mod 5 irreducibility test.

15. If a regular 9-gon is constructible, then so is the angle $360°/9 = 40°$. But Exercise 10 shows that a $40°$ angle is not constructible.

16. Use Exercises 5, 6, and 7.

17. This amounts to showing $\sqrt{\pi}$ is not constructible. But if $\sqrt{\pi}$ is constructible, so is π. However, $[Q(\pi) : Q]$ is infinite.

18. It suffices to show that $2\pi/5$ is constructible. Clearly, $[Q(\cos 2\pi/5) : Q] = 2$ so that $\cos 2\pi/5$ is constructible. Now use Exercise 6.

19. "Tripling" the cube is equivalent to constructing an edge of length $\sqrt[3]{3}$. But $[Q(\sqrt[3]{3}) : Q] = 3$, so this can't be done. No, for $4V$ since $[Q(\sqrt[3]{4}) : Q] = 3$. Yes, for $8V$ since $[Q(\sqrt[3]{8}) : Q] = 1$, which is a power of 2.

20. "Cubing" the circle is equivalent to constructing the length $\sqrt[3]{\pi}$. But $[Q(\sqrt[3]{\pi}) : Q]$ is infinite.

21. Use Exercises 1-4.

CHAPTER 23

Sylow Theorems

1. $a = eae^{-1}$; $cac^{-1} = b$ implies $a = c^{-1}bc = c^{-1}b(c^{-1})^{-1}$; $a = xbx^{-1}$ and $b = ycy^{-1}$ imply $a = xycy^{-1}x^{-1} = xyc(xy)^{-1}$.

2. Observe that $bab^{-1} = \phi_b(a)$ where ϕ_b is the inner automorphism induced by b.

3. Note that $|a^2| = |a|/2$ and appeal to Exercise 2.

4. $\{e\}, \{a^2\}, \{a, a^3\}, \{b, ba^2\}, \{ba, ba^3\}$

5. Observe that $T(xC(a)) = xax^{-1} = yay^{-1} = T(yC(a)) \Leftrightarrow y^{-1}xa = ay^{-1}x \Leftrightarrow y^{-1}x \in C(a) \Leftrightarrow yC(a) = xC(a)$. This proves that T is well defined and one-to-one. Onto is by definition.

6. $\text{cl}(a) = \{a\}$ if and only if for all x in G, $xax^{-1} = a$. This is equivalent to $a \in Z(G)$.

7. Say $\text{cl}(e)$ and $\text{cl}(a)$ are the only two conjugacy classes of a group G of order n. Then $\text{cl}(a)$ has $n - 1$ elements and, by Theorem 24.1 $|C(a))| = n - 1$. Because $C(a)$ is a subgroup of G, we have that $n - 1$ divides n. But the only time $n - 1$ divides n is when $n = 2$.

8. By Sylow's Third Theorem the number of Sylow 7 subgroups is 1 or 8. So the number of elements of order 7 is 6 or 48.

9. It suffices to show that the correspondence from the set of left cosets of $N(H)$ in G to the set of conjugates of H given by $T(xN(H)) = xHx^{-1}$ is well defined, onto, and one-to-one. Observe that $xN(H) = yN(H) \Leftrightarrow y^{-1}xN(H) = N(H) \Leftrightarrow y^{-1}x \in N(H) \Leftrightarrow y^{-1}xH(y^{-1}x)^{-1} = y^{-1}xHx^{-1}y = H \Leftrightarrow xHx^{-1} = yHy^{-1}$. This shows that T is well defined and one-to-one. By observation, T is onto.

10. Let r denote the number of conjugates of H. By Exercise 9 we have $r = |G : N(H)|$. Since each conjugate of H has the same order as H and contains the identity and $H \subseteq N(H)$, we know that the union of all the conjugates of H has fewer than $|G : N(H)||H| \leq |G : H||H| = |G|$ elements because the identity occurs r times in the union.

11. Say $\text{cl}(x) = \{x, g_1xg_1^{-1}, g_2xg_2^{-1}, \ldots, g_kxg_k^{-1}\}$. If $x^{-1} = g_ixg_i^{-1}$, then for each $g_jxg_j^{-1}$ in $\text{cl}(x)$ we have $(g_jxg_j^{-1})^{-1} = g_jx^{-1}g_j^{-1} = g_j(g_ixg_i^{-1})g_j^{-1} \in \text{cl}(x)$. Because $|G|$ has odd order, $g_jxg_j^{-1} \neq (g_jxg_j^{-1})^{-1}$. It follows that $|\text{cl}(x)|$ is even. But this contradicts the fact that $|\text{cl}(x)|$ divides $|G|$.

12. By Theorem 9.3, we know that in each case the center of the group is the identity. So, in both cases the first summand is 1. In the case of 39, all the summands after the first one must be 3 or 13. In the case of 55, all the summands after the first one must be 5 or 11. Thus the only possible class equations are $39 = 1 + 3 + 3 + 3 + 3 + 13 + 13$; $55 = 1 + 5 + 5 + 11 + 11 + 11 + 11$.

13. Part **a** is not possible by the Corollary of Theorem 23.2. Part **b** is not possible because it implies that the center would have order 2 and 2 does not divide 21. Part **c** is the class equation for D_5. Part **d** is not possible because of Corollary 1 of Theorem 23.1.

14. Since $Z(G) = \{R_0, R_{180}\}$ we have two occurrences of 1 in the class equation.

15. Let H and K be distinct Sylow 2-subgroups of G. By Theorem 7.2, we have $48 \geq |HK| = |H||K|/|H \cap K| = 16 \cdot 16/|H \cap K|$. This simplifies to $|H \cap K| > 5$. Since H and K are distinct and $|H \cap K|$ divides 16, we have $|H \cap K| = 8$.

16. This is immediate by induction.

17. By Example 5 of Chapter 9, $\langle x \rangle K$ is a subgroup. By Theorem 7.2, $|\langle x \rangle K| = |\langle x \rangle||K|/|\langle x \rangle \cap K|$. Since K is a Sylow p-subgroup it follows that $\langle x \rangle = \langle x \rangle \cap K$. Thus $\langle x \rangle \subseteq K$.

18. $aba^2b = a(ba)ab = a(a^2b)ab = a^3(ba)b = a^3(a^2b)b = a^5b^2$.

19. By Theorem 23.5, n_p, the number of Sylow p-subgroups has the form $1 + kp$ and n_p divides $|G|$. But if $k \geq 1$, $1 + kp$ is relatively prime to p^n and does not divide m. Thus $k = 0$. Now use the corollary to Theorem 23.5.

20. Use Exercise 17.

21. r does not divide st; s does not divide rt; and t does not divide rs.

22. By Sylow, $n_7 = 1$ or 8. If $n_7 = 8$, the Sylow 7-subgroups contain 48 elements of order 7. This means all the elements whose orders are a power of 2 belong to a single Sylow 2-subgroup. So $n_2 = 1$.

23. There are two Abelian groups of order 4 and two of order 9. There are both cyclic and dihedral groups of orders 6, 8, 10, 12, and 14. So, 15 is the first candidate. And, in fact, Theorem 23.5 shows that there is only one group of order 15.

24. $n_3 = 7$, otherwise the group is the internal direct product of subgroups of orders 3 and 7 and such a group is cyclic.

25. Let n_p denote the number of Sylow p-subgroups of G. By Sylow's Third Theorem, $n_p = 1, q$ or q^2. Clearly, p does not divide $q - 1$, so $n_p \neq q$. If p is a divisor of $q^2 - 1 = (q-1)(q+1)$, then p divides $q - 1$ or $q + 1$. But $p > q$ implies $p \geq q + 2 > q + 1 > q - 1$. So, $n_p \neq q^2$. Thus, $n_p = 1$.

26. Let H be a subgroup of order 3 and K a subgroup of order 5 in one of the six groups. If one of H or K is normal in the whole group then HK is a cyclic subgroup (Theorem 23.6) of order 15. But a permutation of order 15 written in disjoint cycle form requires at least one 3-cycle and one 5-cycle.

27. A group of order 100 has 1, 5 or 25 subgroups of order 4; exactly one subgroup of order 25 (which is normal); at least one subgroup of order 5; and at least one subgroup of order 2.

28. Mimic Example 6.

29. Let H, K, and L be Sylow 5-, 7-, and 17- subgroups of G, respectively. By the Sylow theorems, H is normal in G and by Example 5 in Chapter 9 and Theorem 23.6, HK and HL are cyclic subgroups of G of orders of 35 and 85. Thus, $|C(H)$ is divisible by 5, 7, and 17. So, $C(H) = G$. This means that H is a subgroup of $Z(G)$.
Alternate solution. Let H be a Sylow 5-subgroup. Since the number of Sylow 5-subgroups is 1 mod 5 and divides $7 \cdot 17$, the only possibility is 1. So, H is normal in G. Then by the N/C Theorem (Example 16 of Chapter 10), $|G/C(H)|$ divides both 4 and $|G|$. Thus $C(H) = G$.

30. $2^{|b|} = 1 \bmod |a|$, $4^{|b|} = 1 \bmod |a|$, $2 \cdot 4 = 1 \bmod |a|$, $2^2 = 4 \bmod |a|$; $11^{|b|} = 1 \bmod |a|$, $16^{|b|} = 1 \bmod |a|$, $11 \cdot 16 = 1 \bmod |a|$, $11^2 = 16 \bmod |a|$.

31. By Theorem 23.6, $G/Z(G)$ would be cyclic and therefore by Theorem 9.3 G would be Abelian. But then $G = Z(G)$.

32. $|H| = 1$ or p where p is a prime. To see this, note that if an element g of G has order pn where p is prime, then $|g^n| = p$. Thus $|H|$ must divide p. So the only case when $|H| = p$ is when G has order p^k where p is prime. For if $|G| = p^k m$ where p does not divide m, then by Sylow's First Theorem, G would have an element of some prime order $q \neq p$. But then $|H|$ would divide both p and q.

33. By Theorem 23.5, n_p has the form $1 + kp$ and divides q. Thus, $k = 0$, and by Corollary 1 of Theorem 23.5, the Sylow p-subgroup of G is normal.

34. By Theorem 23.4 the groups have subgroups H and K of orders 3 and 5, respectively. By Theorem 23.5 n_1 is 1 or 25. If $n_3 = 1$, then $|HK| = 15$. (See Example 5 of Chapter 9.) If $n_3 = 25$, then by Exercise 9 $|G : N(H)| = 25$ and therefore $|N(H)| = 15$.

35. Sylow's Third Theorem (Theorem 23.5) implies that the Sylow 3- and Sylow 5-subgroups are unique. Pick any x not in the union of these. Then $|x| = 15$.

36. By Sylow, $n_7 = 1$ or 15, and $n_5 = 1$ or 21. Counting elements reveals that at least one of these must be 1. Let H be a Sylow-7 subgroup and K be a Sylow-5 subgroup. Then HK is a subgroup of order 35 and is cyclic by Theorem 23.6. So, HK has $\phi(5)\phi(7) = 24$ elements of order 35. If H is normal, then G has $21 \cdot 4 = 84$ elements of order 5. This gives at least 108 elements in G. If K is normal, then G has $15 \cdot 6 = 90$ elements of order 7. This gives at least 114 elements in G.

37. By Sylow's Third Theorem, $n_{17} = 1$ or 35. Assume $n_{17} = 35$. Then the union of the Sylow 17-subgroups has 561 elements. By Sylow's Third Theorem, $n_5 = 1$. Thus, we may form a cyclic subgroup of order 85 (Example 5 of Chapter 9 and Theorem 23.6). But then there are 64 elements of order 85. This gives too many elements for the group.

38. By Sylow, $n_5 = 1$ or 6. A_5 has 24 elements of order 5.

39. If $|G| = 60$ and $|Z(G)| = 4$, then by Theorem 23.6, $G/Z(G)$ is cyclic. The "G/Z" Theorem (Theorem 9.3) then tells us that G is Abelian. But if G is Abelian, then $Z(G) = G$.

40. a. Form a factor group G/N of order 30. The discussion preceding Theorem 23.6 shows G/N has normal subgroups of orders 3, 5, and 15. Now pullback.
    b. Let H be a Sylow 2-subgroup containing N. The product of H with the subgroups of orders 6 and 10 have orders 12 and 20.

c. The product of N and the subgroup of order 15 has order 30 and is an internal direct product.

41. Let H be the Sylow 3-subgroup and suppose that the Sylow 5-subgroups are not normal. By Sylow's Third Theorem, there must be six Sylow 5-subgroups, call them $K_1, \ldots, K_6$. These subgroups have 24 elements of order 5. Also, the cyclic subgroups $HK_1, \ldots, HK_6$ of order 15 each have eight generators. Thus, there are 48 elements of order 15. This gives us more than 60 elements in G.

42. Let N be the normal subgroup of order 4. Then by Sylow, G/N has a normal Sylow 7-subgroup whose pullback is normal.

43. We proceed by induction on $|G|$. By Theorem 23.2 and Theorem 9.5, $Z(G)$ has an element x of order p. By induction, the group $G/\langle x\rangle$ has normal subgroups of order p^k for every k between 1 and $n-1$, inclusively. By Exercise 51 in Chapter 10 and Exercise 59 of Chapter 9, every normal subgroup of $G/\langle x\rangle$ has the form $H/\langle x\rangle$, where H is a normal subgroup of G. Moreover, if $|H/\langle x\rangle| = p^k$, then $|H|$ has order p^{k+1}.

44. Let $x \in G$ have maximum order, $|x| = p^t$. Now let y belong to G. Then $|y| = p^s \leq p^t$. Since $\langle x\rangle$ has a subgroup of order p^s, we have $\langle y\rangle \subseteq \langle x\rangle$.
Alternate solution. Suppose that G is not cyclic. Then every element of G belongs to a subgroup of order less than p^n. But summing the number of elements in the n proper subgroups we get less that $1+p+p^2+\cdots+p^{n-1} < p^n$, which is a contradiction

45. Pick $x \in Z(G)$ such that $|x| = p$. If $x \notin H$, then $x \in N(H)$ and we are done. If $x \in H$, by induction, $N(H/\langle x\rangle) > H/\langle x\rangle$, say $y\langle x\rangle \in N(H/\langle x\rangle)$ but $y\langle x\rangle \notin H/\langle x\rangle$. Then $y \notin H$ and for any $h \in H$ we have $yhy^{-1}\langle x\rangle = y\langle x\rangle h\langle x\rangle y^{-1}\langle x\rangle \in H/\langle x\rangle$. So, $yhy^{-1} \in H$ and therefore $y \in N(H)$.

46. Since both H and xHx^{-1} have the same set of conjugates this statement follows directly from Exercise 9.

47. By Sylow's Second Theorem, K is contained in some Sylow p-subgroup H. Then for any x in G we have $K = xKx^{-1} \subset xHx^{-1}$. By Sylow's Theorem every Sylow p-subgroup has the form xHx^{-1}.
Alternate proof. Let K be a Sylow p-subgroup. Then by Example 5 of Chapter 9 HK is a subgroup of G and by Theorem 7.2 $|HK| = |H||K|/|H \cap K|$. Since this is a power of p, we must have $|HK| = |K|$ and because K is contained in HK, we have $HK = K$. Thus H is contained in K.

48. If there is only one Sylow 2-subgroup it is normal by the corollary of Theorem 23.5. Otherwise, let H and K be distinct Sylow 2-subgroups. Then, by Lagrange, $|H \cap K| = 1, 2, 4$, or 8. If $|H \cap K| = 1, 2$ or 4 then the set HK has at least $|H||K|/|H \cap K| \geq 16 \cdot 16/4 = 64$, which is impossible in a group of order 48. If $|H \cap K| = 8$, then because $H \cap K$ is a subgroup of both H and K, by Exercise 45, $N(H \cap K)$ contains both H and K. But then 16 divides $|N(H \cap K)|$ and $N(H \cap K)$ has more than 16 elements. So $N(H \cap K)$ is the entire group and therefore $H \cap K$ is normal in G.

49. Sylow's Third Theorem shows that all the Sylow subgroups are normal. Then Theorem 7.2 and Example 5 of Chapter 9 ensure that G is the internal direct product of its Sylow subgroups. G is cyclic because of Theorems 9.6 and 8.2. G is Abelian because of Theorem 9.6 and Exercise 4 in Chapter 8.

50. Let $p_1^{n_1} p_2^{n_2} \cdots p_k^{n_k}$ be the prime power decomposition of $|G|$. For each p_i let S_p be the unique Sylow p-subgroup of G. By Corollary 1 of Theorem 23.5 and Exercise 44 we know that each S_p is normal in G and cyclic. It follows from Corollary 1 of Theorem 8.2 and Theorem 9.6 that G is cyclic.

51. Since automorphisms preserve order, we know $|\alpha(H)| = |H|$. But then the corollary of Theorem 23.5 shows that $\alpha(H) = H$.

52. Clearly, $N(H) \subseteq N(N(H))$. Let x belong to $N(N(H))$. Since $H \subseteq N(H)$, for any h in H we have that xhx^{-1} belongs to $N(H)$. By Theorem 7.2, $|\langle xhx^{-1}\rangle H| = |\langle xhx^{-1}\rangle||H|/|\langle xHx^{-1}\rangle \cap H|$. Since $|\langle xhx^{-1}\rangle|$ is a power of p and H is a Sylow p-subgroup we must have $\langle xhx^{-1}\rangle = \langle xhx^{-1}\rangle \cap H \subseteq H$. Thus, xhx^{-1} is in H and x is in $N(H)$.

53. That $|N(H)| = |N(K)|$ follows directly from the last part of Sylow's Third Theorem and Exercise 9.

54. Mimic Example 6. Three pairs are: 5, 7; 7, 11; 11, 13.

55. Normality of H implies $\text{cl}(h) \subseteq H$ for h in H. Thus the conjugacy classes of H obtained by conjugating by elements from G are subsets of H. Moreover, since every element h in H is in $\text{cl}(h)$ the union of the conjugacy classes of H is H. This is true only when H is normal.

56. By Sylow's Second Theorem we know H is a subgroup of a Sylow p-subgroup of G, call it K. If H is a proper subgroup of K, then by Exercise 43 there is an element a in the normalizer of H restricted to K that is not in H. Then $H\langle a\rangle$ is a p-subgroup of $N(H)$ of order larger than H. But then H is not a Sylow p-subgroup of $N(H)$.

57. Suppose that G is a group of order 12 that has nine elements of order 2. By the Sylow Theorems, G has three Sylow 2-subgroups whose union contains the identity and the nine elements of order 2. If H and K are both Sylow 2-subgroups, by Theorem 7.2 $|H \cap K| = 2$. Thus the union of the three Sylow 2-subgroups has at most 7 elements of order 2 since there are 3 in H, 2 more in K that are not in H, and at most 2 more that are in the third but not in H or K.

58. By way of contradiction, assume that H is the only Sylow 2-subgroup of G and that K is the only Sylow 3-subgroup of G. Then H and K are normal and Abelian (corollary to Theorem 23.5 and corollary to Theorem 23.2). So, $G = H \times K \approx H \oplus K$ and, from Exercise 4 of Chapter 8, G is Abelian.

59. By Lagrange's Theorem, any nontrivial proper subgroup of G has order p or q. It follows from Theorem 23.5 and its corollary that there is exactly one subgroup of order q which is normal (for otherwise there would be $(q+1)(q-1) = q^2 - 1$ elements of order q). On the other hand, there cannot be a normal subgroup of order p, for then G would be an internal direct product of a cyclic group of q and a cyclic group of order p, which is Abelian. So, by Theorem 23.5 there must be exactly q subgroups of order p.

60. Mimic Example 6.

61. Note that any subgroup of order 4 in a group of order $4m$ where m is odd is a Sylow 2-subgroup. By Sylow's Third Theorem, the Sylow 2-subgroups are conjugate and therefore isomorphic. S_4 contains both the subgroups $\langle(1234)\rangle$ and $\{(1), (12), (34), (12)(34)\}$.

62. By the "G/Z Theorem" (9.3), $|N(H)/C(H)|$ divides $|\mathrm{Aut}(H)|$. Since H is cyclic we know that $C(H) \supseteq H$ and therefore $|N(H)/C(H)|$ is relatively prime to p. Letting $|H| = p^k$, we have by Theorem 6.5 that $\mathrm{Aut}(H)$ is isomorphic to $U(p^k)$ and by the formula given in Chapter 8 we have $|U(p^k)| = p^{k-1}(p-1)$. Since the smallest prime divisor of $|N(H)/C(H)|$ is greater than p, we must have $|N(H)/C(H)| = 1$.

63. By Sylow's Third Theorem, the number of Sylow 13-subgroups is equal to 1 mod 13 and divides 55. This means that there is only one Sylow 13-subgroup, so it is normal in G. Thus $|N(H)/C(H)| = 715/|C(H)|$ divides both 55 and 12. This forces $715/|C(H)| = 1$ and therefore $C(H) = G$. This proves that H is contained in $Z(G)$. Applying the same argument to K we get that K is normal in G and $|N(K)/C(K)| = 715/|C(K)|$ divides both 65 and 10. This forces $715/|C(K)| = 1$ or 5. In the latter case K is not contained in $Z(G)$.

64. In each case the number of Sylow 3-subgroups is the number of elements of order 3 divided by 2 since a group of order 3 has two elements of order 3 and no two of subgroups of order 3 share an element of ordder 3. So, In S_4 there are $4 \cdot 3 \cdot 2/3 = 8$ elements of order 3 and 4 subgroups of order 3; in S_5 there are $5 \cdot 4 \cdot 3/3 = 20$ elements of order 3 and 10 subgroups of order 3;

65. Using Sylow's third theorem and counting the number of elements of orders 2, 3, and 5, we have that the number of Sylow p-subgroups in each respective cases is 15, 10, and 6.

66. By the previous exercise and Corollary 2 of Theorem 23.5, we know that $|N(H)| = 10$. Also, by the classification of groups of order $2p$, where p is a prime (Theorem 7.3), we know that $N(H)$ is isomorphic to Z_{10} or D_5. However, A_5 has no element of order 10.

67. One of many is (12)(45).

CHAPTER 24

Finite Simple Groups

1. This follows directly from the "2·odd" Theorem (Theorem 24.2).

2. We may assume that $n_5 = 56$ and $n_7 = 8$. Then counting elements forces $n_2 = 1$. Alternate proof. From $n_5 = 56$ and $n_7 = 8$, we have that for any Sylow 7-subgroup L_7, $N|(L_7)| = 35$. Because $N(L_7)$ is a cyclic group, it has a Sylow 5-subgroup L_5 whose normalize $N(L_5)$ contains $N(L_7)$. But $n_5 = 56$ implies that $|N(L_5)| = 5$.

3. By the Sylow Theorems, if there were a simple group of order 216 the number of Sylow 3-subgroups would be 4. Then the normalizer of a Sylow 3-subgroup would have index 4. The Index Theorem (corollary of Theorem 24.3) then gives a contradiction.

4. Observe that $n_5 = 6$ and use the Index Theorem.

5. Suppose G is a simple group of order 525. Let L_7 be a Sylow 7-subgroup of G. It follows from Sylow's theorems that $|N(L_7)| = 35$. Let L be a subgroup of $N(L_7)$ of order 5. Since $N(L_7)$ is cyclic (Theorem 24.6), $N(L) \geq N(L_7)$, so that 35 divides $|N(L)|$. But L is contained in a Sylow 5-subgroup (Theorem 23.4), which is Abelian (see the Corollary to Theorem 23.2). Thus, 25 divides $|N(L)|$ as well. It follows that 175 divides $|N(L)|$. The Index Theorem now yields a contradiction.

6. The Index Theorem rules out $n_5 = 6$. So we may assume that $n_3 = 7$ and $n_5 = 21$. Let L_5 be a Sylow 5-subgroup. Then $n_5 = 21$ implies that $|N(L_5)| = 15$. By Theorem 23.6, $N(L_5)$ has an element of order 15 and by the Embedding Theorem (Corollary 2 of Theorem 24.3), G is isomorphic to a subgroup of A_7. But A_7 does not have an element of order 15 because any such element written in disjoint cycle form requires at least one 3-cycle and one 5-cycle.

7. Suppose that there is a simple group G of order 528 and L_{11} is a Sylow 11-subgroup. Then $n_{11} = 12$, $|N(L_{11})| = 44$, and G is isomorphic to a subgroup of A_{12}. Then by the N/C Theorem in Example 17 of Chapter 10, $|N(L_{11})/C(L_{11})|$ divides $|\text{Aut}(Z_{11})| = 10$, $|C(L_{11})| = 22$ or 44. In either case, $C(L_{11})$ has elements of order 2 and 11 that commute. But then $C(L_{11})$ has an element of order 22 whereas A_{12} does not.

8. We may assume the $n_3 = 10$ and $n_5 = 6$ or 36. The Index Theorem rules of 6. Let L_5 be a Sylow 5-subgroup. This implies that $|N(L_5)| = 15$. By Theorem 23.6, we know that $N(L_5)$ is cyclic. Let K be the subgroup of $N(L_5)$ of order 3 and let L_3 be a Sylow subgroup of G that contains K (see Sylow's Second Theorem). Then L_5K is a cyclic subgroup of $N(K)$ of order 15. So, 15 divides $|N(K)|$. Moreover, by Exercise 45 of Chapter 23 and Sylow's First Theorem, $N(K)$ contains a subgroup K' of L_3 of order 9. Thus L_5K' is a group of order 45. Because groups of order 45 are Abelian, (see Example 6 in Chapter 23), $N(L_5)$ has order at least 45. But $n_5 = 36$ means $|N(L_5)| = 15$.

9. Suppose that there is a simple group G of order 396 and L_{11} is a Sylow 11-subgroup. Then $n_{11} = 12$, $|N(L_{11})| = 33$, and G is isomorphic to a subgroup of A_{12}. Since

$|N(L_{11})/C(L_{11})|$ divides $|\text{Aut}(Z_{11})| = 10$, $|C(L_{11})| = 33$. Then $C(L_{11})$ has elements of order 3 and 11 that commute. But then $C(L_{11})$ has an element of order 33 whereas A_{12} does not.

10. 211, 223, 227, 229 and 233 are prime. The 2·odd test rules out 202, 206, 210, 214, 218, 222, 226, 230 and 234. The Index Theorem rules out 216 and 224. The Sylow test rules out the remaining cases.

11. If we can find a pair of distinct Sylow 2-subgroups A and B such that $|A \cap B| = 8$, then $|N(A \cap B) \geq AB|$, so that $N(A \cap B) = G$. Now let H and K be any pair of distinct Sylow 2-subgroups. Then $16 \cdot 16/|H \cap K| = |HK| \leq 112$ (Theorem 7.2), so that $|H \cap K|$ is at least 4. If $|H \cap K| = 8$, we are done. So, assume $|H \cap K| = 4$. Then $N(H \cap K)$ picks up at least 8 elements from H and at least 8 from K (see Exercise 45 of Chapter 23). Thus, $|N(H \cap K)| \geq 16$ and is divisible by 8. So, $|N(H \cap K)| = 16$, 56, or 112. Since the latter two cases imply that G has a normal subgroup, we may assume $|N(H \cap K)| = 16$. If $N(H \cap K) = H$, then $|H \cap K| = 8$, since $N(H \cap K)$ contains at least 8 elements from K. So, we may assume that $N(H \cap K) \neq H$. Then, we may take $A = N(H \cap K)$ and $B = H$.

12. Suppose that H is a subgroup of S_n of order $n!/2$ other than A_n. Then $A_nH = S_n$ and it follows from Theorem 7.2 that $A_n \cap H$ is a subgroup of A_n of index 2. But a subgroup of index 2 is normal (see Exercise 9 in Chapter 9).

13. If H is a proper subgroup of A_{n+1} of order greater than $n!/2$, then $[A_{n+1} : H] < [A_{n+1} : A_n] = n + 1$ and it follows from the Embedding Theorem that A_{n+1} is isomorphic to a subgroup of A_n, which is impossible.

14. Theorem 23.2 handles the case where $p = q = r$. The Sylow Test for nonsimplicity (Theorem 24.1) shows that a group of order p^2q with $p > q$ must have a normal Sylow p-subgroup. The same theorem implies that a simple group of order p^2q with $p < q$ would have p^2 subgroups of order q and more than one subgroup of order p^2. Counting elements then yields a contradiction.

    Finally consider a simple group of order pqr where $p < q < r$. Then there are pq subgroups of order r, at least r subgroups of order q and at least q subgroups of order p. But $pq(r-1) + r(q-1) + q(p-1) > pqr$.

15. If A_5 had a subgroup of order 30, 20, or 15, then there would be a subgroup of index 2, 3 or 4. But then the Index Theorem gives us a contradiction to the fact that G is simple.

16. The argument for Exercise 15 applies here as well.

17. (Solution by Gurmeet Singh) By Sylow's Third Theorem we know that number of Sylow 5-subgroups is 6. This means that 6 is the index of the normalizer of a Sylow 5-subgroup. But then, by the Embedding Theorem, G is isomorphic to a subgroup of order 120 in A_6. This contradicts Exercise 16.

18. Use Exercise 51 of Chapter 9.

19. Let α be as in the proof of the Generalized Cayley Theorem (Theorem 24.3). Then, if $g \in \text{Ker}\ \alpha$ we have $gH = T_g(H) = H$ so that $\text{Ker}\ \alpha \subseteq H$. Since $\alpha(G)$ consists of a group of permutations of the left cosets of H in G, we know by the First Isomorphism Theorem (Theorem 10.3) that $G/\text{Ker}\ \alpha$ is isomorphic to a subgroup of $S_{|G:H|}$. Thus, $|G/\text{Ker}\ \alpha|$ divides $|G : H|!$. Since $\text{Ker}\ \alpha \subseteq H$, we have that

$|G : H||H : \text{Ker } \alpha| = |G : \text{Ker } \alpha|$ must divide $|G : H|! = |G : H|(|G : H| - 1)!$. Thus, $|H : \text{Ker } \alpha|$ divides $(|G : H| - 1)!$. Since $|H|$ and $(|G : H| - 1)!$ are relatively prime, we have $|H : \text{Ker } \alpha| = 1$ and therefore $H = \text{Ker } \alpha$. So, by the Corollary of Theorem 10.2, H is normal. In the case that a subgroup H has index 2, we conclude that H is normal.

20. If ϕ is such a homomorphism then by Theorem 10.3 $G/\text{Ker } \phi \approx S_n$. Since $|G| > n!$ we must have $|\text{Ker } \phi| > 1$. So, Ker ϕ is a proper, non-trivial normal subgroup of G.

21. If H is a proper normal subgroup of S_5, then $H \cap A_5 = A_5$ or $\{\varepsilon\}$ since A_5 is simple and $H \cap A_5$ is normal. But $H \cap A_5 = A_5$ implies $H = A_5$, whereas $H \cap A_5 = \{\varepsilon\}$ implies $H = \{\varepsilon\}$ or $|H| = 2$ because in a subgroup of S_n every element is even or half of its elements are even. (See Exercise 27 of Chapter 5.) If $|H| = 2$, then it has an element of the form (ab). But for any $c \in \{1, 2, \ldots, 5\}$ other than a or b, $(ac)(ab)(ac) \neq (ab)$. So this case is ruled out.

22. The Sylow Test for Nonsimplicity yields $n_5 = 6$ and $n_3 = 4$ or 10. The Index Theorem rules out $n_3 = 4$. Let L_3 be the Sylow 3-subgroup of the group and L_5 be the Sylow 5-subgroup of the group. Then, by Corollary 1 of Theorem 24.3, we have $|N(L_3| = 6$ and $|N(L_5)| = 10$.

23. From Table 5.1 we see that the Sylow 2-subgroup of A_4 is unique and therefore normal in A_4.

24. Let $H = \langle (12), (12345) \rangle$. First note that $(12345)(12)(12345)^{-1} = (23)$ implies that $(12)(12345)(12)(12345)^{-1} = (12)(23) = (123)$ so H contains an element of order 3. Moreover, $(12345)(12) = (1345) \in H$. This means that $|H|$ is divisible by 3, 4, and 5 and therefore $|H| = 60$ or 120. But $|H|$ cannot be 60 for if so, then the subset of even permutations in H would be a subgroup of order 30 (see Exercise 27 in Chapter 5). This means that A_5 would have a subgroup of index 2, which would a normal subgroup. This contradicts the simplicity of A_5.

25. Suppose that S_5 has a subgroup H that contains a 5-cycle α and a 2-cycle β. Say $\beta = (a_1a_2)$. Then there is some integer k such that $\alpha^k = (a_1a_2a_3a_4a_5)$. Note that $(a_1a_2a_3a_4a_5)^{-1}(a_1a_2)(a_1a_2a_3a_4a_5)(a_1a_2) = (a_5a_4a_3a_2a_1)(a_1a_2)(a_1a_2a_3a_4a_5)(a_1a_2) = (a_1a_2a_5)$, so H contains an element of order 3. Moreover, since $\alpha^{-2}\beta\alpha^2 = (a_4a_2a_5a_3a_1)(a_1a_2)(a_1a_3a_5a_2a_4) = (a_4a_5)$, H contains the subgroup $\{(1), (a_1a_2), (a_4a_5), (a_1a_2)(a_4a_5)\}$. This means that $|H|$ is divisible by 60. But $|H|$ cannot be 60 for if so, then the subset of even permutations in H would be a subgroup of order 30 (see Exercise 27 in Chapter 5). This means that A_5 would have a subgroup of index 2, which would be a normal subgroup. This contradicts the simplicity of A_5.

26. Let $p = |G : H|$ and $q = |G : K|$. It suffices to show that $p = q$. But if $p < q$, say, then q does not divide $p!$. This contradicts the Index Theorem.

27. Suppose there is a simple group of order 60 that is not isomorphic to A_5. The Index Theorem implies $n_2 \neq 1$ or 3, and the Embedding Theorem implies $n_2 \neq 5$. Thus, $n_2 = 15$. If every pair of Sylow 2-subgroups has only the identity element in common, then the union of the 15 Sylow 2-subgroups has 46 elements. But $n_5 = 6$, so there are also 24 elements of order 5. This gives more than 60. As was the case in showing that there is no simple group of order 144, the normalizer of this intersection has index 5, 3, or 1. But the Embedding Theorem and the Index Theorem rule these out.

28. Say G is a simple group with a subgroup of index 4. By the Embedding Theorem G is isomorphic to a subgroup of A_4. Since $|G|$ is divisible by 4, we must have that $|G| = 4$ or 12. The subgroup of A_4 of order 4 is isomorphic to $Z_2 \times Z_2$ so it is not simple. Since the Sylow 2-subgroup of A_4 is the unique subgroup of order 4 it is normal in A_4, A_4 is not simple.

29. Suppose there is a simple group G of order p^2q where p and q are odd primes and $q > p$. Since the number of Sylow q-subgroups is 1 mod q and divides p^2, it must be p^2. Thus there are $p^2(q-1)$ elements of order q in G. These elements, together with the p^2 elements in one Sylow p-subgroup, account for all p^2q elements in G. Thus there cannot be another Sylow p-subgroup. But then the Sylow p-subgroup is normal in G.

30. Let L and M be distinct maximal subgroups of G. Since $N(K)$ contains both L and M it properly contains them. Since L is maximal, $N(K) = G$. Because G is simple, we have $K = \{e\}$.

31. Consider the right regular representation of G. Let g be a generator of the Sylow 2-subgroup and suppose that $|G| = 2^k n$ where n is odd. Then every cycle of the permutation T_g in the right regular representation of G has length 2^k. This means that there are exactly n such cycles. Since each cycle is odd and there is an odd number of them, T_g is odd. This means that the set of even permutations in the regular representation has index 2 and is therefore normal. (See Exercise 27 in Chapter 5 and Exercise 9 in Chapter 9.)

32. If S_5 had a subgroup H of order 40 or 30, then $H \cap S_5$ would be a subgroup of A_5 of order 20, 30 or 15 since every element of H is even or half of them are even (40 is excluded by Lagrange's Theorem). By Exercise 15, this is impossible.

33. If ϕ is such a homomorphism, then by Theorem 10.3, $G/\text{Ker } \phi \approx S_n$. Since $|G| > n!$ we must have $|\text{Ker } \phi| > 1$. So, Ker ϕ is a proper, non-trivial normal subgroup of G.

34. See Exercise 6.

CHAPTER 25

Generators and Relations

1. $u \sim u$ because u is obtained from itself by no insertions; if v can be obtained from u by inserting or deleting words of the form xx^{-1} or $x^{-1}x$, then u can be obtained from v by reversing the procedure; if u can be obtained from v and v can be obtained from w, then u can be obtained from w by first obtaining v from w then u from v.

2. Let a be any reflection in D_n and let $b = aR_{360/n}$. Then $aZ(D_n)$ and $bZ(D_n)$ have order 2 and generate $D_n/Z(D_n)$. Now use Theorem 25.5 and the fact that $|D_n/Z(D_n)| = n = |D_{n/2}|$.

3. $$\begin{aligned} b(a^2N) &= b(aN)a = (ba)Na = a^3bNa = a^3b(aN) = a^3(ba)N \\ &= a^3a^3bN = a^6bN = a^6Nb = a^2Nb = a^2bN \\ b(a^3N) &= b(a^2N)a = a^2bNa = a^2b(aN) = a^2a^3bN \\ &= a^5bN = a^5Nb = aNb = abN \\ b(bN) &= b^2N = N \\ b(abN) &= baNb = a^3bNb = a^3b^2N = a^3N \\ b(a^2bN) &= ba^2Nb = a^2bNb = a^2b^2N = a^2N \\ b(a^3bN) &= ba^3Nb = abNb = ab^2N = aN \end{aligned}$$

4. Since $b = b^{-1}$, we have $bab = a^2$. Then $a = a^6 = (bab)^3 = ba^3b$ so that $ba = a^3b$. Thus, $a^3b = a^2b$ and $a = e$. Finally, note that Z_2 satisfies the relations with $a = 0$ and $b = 1$.

5. Let F be the free group on $\{a_1, a_2, \ldots, a_n\}$. Let N be the smallest normal group containing $\{w_1, w_2, \ldots, w_t\}$ and let M be the smallest normal subgroup containing $\{w_1, w_2, \ldots, w_t, w_{t+1}, \ldots, w_{t+k}\}$. Then $F/N \approx G$ and $F/M \approx \overline{G}$. The homomorphism from F/N to F/M given by $aN \to aM$ induces a homomorphism from G onto $\bar{G}$.

   To prove the corollary, observe that the theorem shows that K is a homomorphic image of G, so that $|K| \leq |G|$.

6. Use the Corollary to Dyck's Theorem.

7. Clearly, a and ab belong to $\langle a, b\rangle$, so $\langle a, ab\rangle \subseteq \langle a, b\rangle$. Also, a and $a^{-1}(ab) = b$ belong to $\langle a, ab\rangle$.

8. Use Theorem 25.5.

9. By Exercise 7, $\langle x, y\rangle = \langle x, xy\rangle$. Also, $(xy)^2 = (xy)(xy) = (xyx)y = y^{-1}y = e$, so by Theorem 25.5, G is isomorphic to a dihedral group and from the proof of Theorem 25.5, $|x(xy)| = |y| = n$ implies that $G \approx D_n$.

10. 3. $\langle x, y, z \mid x^2 = y^2 = z^2 = e, xy = yx, xz = zx, yz = zy\rangle$.

11. Since $x^2 = y^2 = e$, we have $(xy)^{-1} = y^{-1}x^{-1} = yx$. Also, $xy = z^{-1}yz$, so that $(xy)^{-1} = (z^{-1}yz)^{-1} = z^{-1}y^{-1}z = z^{-1}yz = xy$.

12. a. $b^0a = a$ b. ba

13. First note that $b^2 = abab$ implies that $b = aba$.

    a. So, $b^2abab^3 = b^2(aba)b^3 = b^2bb^3 = b^6$.

    b. Also, $b^3abab^3a = b^3(aba)b^3a = b^3bb^3a = b^7a$.

14. Observe that G is generated by 6 and 4 where $|6| = 2$, $|4| = 2$ and $|6 \cdot 4| = |3| = n$. Now use Theorem 26.5.

15. First observe that since $xy = (xy)^3(xy)^4 = (xy)^7 = (xy)^4(xy)^3 = yx$, x and y commute. Also, since $y = (xy)^4 = (xy)^3xy = x(xy) = x^2y$ we know that $x^2 = e$. Then $y = (xy)^4 = x^4y^4 = y^4$ and therefore, $y^3 = e$. This shows that $|G| \leq 6$. But Z_6 satisfies the defining relations with $x = 3$ and $y = 2$. So, $G \approx Z_6$.

16. In H, $|xy| = 6$.

17. Note that $yxyx^3 = e$ implies that $yxy^{-1} = x^5$ and therefore $\langle x \rangle$ is normal. So, $G = \langle x \rangle \cup y\langle x \rangle$ and $|G| \leq 16$. From $y^2 = e$ and $yxyx^3 = e$, we obtain $yxy^{-1} = x^{-3}$. So, $yx^2y^{-1} = yxy^{-1}yxy^{-1} = x^{-6} = x^2$. Thus, $x^2 \in Z(G)$. On the other hand, G is not Abelian for if so we would have $e = yxyx^3 = x^4$ and then $|G| \leq 8$. It now follows from the "G/Z" Theorem (Theorem 9.3) that $|Z(G)| \neq 8$. Thus, $Z(G) = \langle x^2 \rangle$. Finally, $(xy)^2 = xyxy = x(yxy) = xx^{-3} = x^{-2}$, so that $|xy| = 8$.

18. Because $\langle a \rangle$ is a subgroup of Q_6 of index 2 it is normal. Thus every element of Q_6 can written in the form a^i or a^ib^j where $0 \leq i < 6$. The relation $b^2 = a^3$ reduces the a^ib^j terms to just a^ib. To see that the only element of order 2 is $a^3 = b^2$ first observe that the relation $b^{-1}ab = a^{-1}$ implies that $b^{-1}a^ib = (b^{-1}ab)^i = a^i$ and therefore $a^ib = ba^{-i}$. Finally, we note that $e = (a^ib)^2 = a^iba^ib = ba^{-i}a^ib = b^2$.

19. Since the mapping from G onto G/N given by $x \to xN$ is a homomorphism, G/N satisfies the relations defining G.

20. If G were Abelian then the relation $st = ts$ could be derived from $sts = tst$. But this same derivation would hold when $s = (23)$ and $t = (13)$. However, $(23)(13) \neq (13)(23)$.

21. For H to be a normal subgroup, we must have $yxy^{-1} \in H = \{e, y^3, y^6, y^9, x, xy^3, xy^6, xy^9\}$. But $yxy^{-1} = yxy^{11} = (yxy)y^{10} = xy^{10}$.

22. Every element has the form x^i or x^iy where $0 \leq i < 2n$. Let $0 < i < 2n$. Then $x^i \in Z(G)$ if and only if $y^{-1}x^iy = x^i$. But

$$y^{-1}x^iy = (y^{-1}xy)^i = (x^{-1})^i = x^{-1}.$$

    So $x^{2i} = e$. This implies $i = n$. A similar argument shows $x^iy \in Z(G)$ implies $i = n$. But $x^ny \in Z(G)$ and $x^n \in Z(G)$ imply $y \in Z(G)$, which is false. So $x^ny \notin Z(G)$. To prove the second portion, observe that $G/Z(G)$ has order $2n$ and is generated by a pair of elements of order 2.

23. First note that $b^{-1}a^2b = (b^{-1}ab)(b^{-1}ab) = a^3a^3 = a^6 = e$. So, $a^2 = e$. Also, $b^{-1}ab = a^3 = a$ implies that a and b commute. Thus, G is generated by an element of order 2 and an element of order 3 that commute. It follows that G is Abelian and has order at most 6. But the defining relations for G are satisfied by Z_6 with $a = 3$ and $b = 2$. So, $G \approx Z_6$.

24. Since $yx = x^3y$, the set $S = \{\langle y\rangle, x\langle y\rangle, x^2\langle y\rangle, x^3\langle y\rangle\}$ is closed under multiplication on the left by x and y. Thus every element of G has the form x^iy^j with $0 \leq i < 4$ and $0 \leq j < 4$.

    To compute the center observe that $xyx = y$ and $xy = yx^3$. So

    $$x^2y = x(xy) = xyx^3 = (xyx)x^2 = yx^2.$$

    Thus $x^2 \in Z(G)$. Also,

    $$xy^2 = (xy)y = yx^3y = yx^2xy = yxyx^2 = y(xyx)x = y^2x$$

    so that $y^2 \in Z(G)$. It follows that $Z(G) = \{e, x^2, y^2, x^2y^2\}$. (Theorem 9.3 shows $|Z(G)| \neq 8$.)

    Finally, observe that $G/\langle y^2\rangle$ has order 8 and is generated by $y\langle y^2\rangle$ and $xy\langle y^2\rangle$ each of which has order 2.

25. In the notation given in the proof of Theorem 25.5 we have that $|e| = 1$, $|a| = |b| = 2$, $|ab| = |ba| = \infty$. Next observe that since every element of D_∞ can be expressed as a string of alternating a's and b's or alternating b's and a's, every element can be expressed in one of four forms: $(ab)^n, (ba)^n, (ab)^na$, or $(ba)^nb$ for some n. Since $|ab| = |ba| = \infty$, we have $|(ab)^n| = |(ba)^n| = \infty$ (excluding $n = 0$). And, since $((ab)^na)^2 = (ab)^na(ab)^na = (ab)(ab)\cdots(ab)a(ab)(ab)\cdots(ab)a$, we can start at the middle and successively cancel the adjacent a's, then adjacent b's, then adjacent a's, and so on to obtain the identity. Thus, $|(ab)^na| = 2$. Similarly, $|(ba)^nb| = 2$.

26. Use Theorem 25.4.

27. First we show that $d = b^{-1}, a = b^2$ and $c = b^3$ so that $G = \langle b\rangle$. To this end, observe that $ab = c$ and $cd = a$ together imply that $cdc = c$ and therefore $d = b^{-1}$. Then $da = b$ and $d = b^{-1}$ together imply that $a = b^2$. Finally, $cd = a$ and $d = b^{-1}$ together imply $c = b^3$. Thus $G = \langle b\rangle$. Now observe that $bc = d, c = b^3$, and $d = b^{-1}$ yield $b^5 = e$. So $|G| = 1$ or 5. But Z_5 satisfies the defining relations with $a = 1, b = 3$, $c = 4$, and $d = 2$.

28. From Theorem 25.5 and its proof, the group is dihedral and has order $2|ab|$. To compute $|ab|$, note that $(ab)^3 = (aba)(bab) = (aba(aba) = aba^2ba = e$. So, the group is D_3.

29. Since $aba^{-1}b^{-1} = e$, G is an Abelian group of order at most 6. Then because Z_6 satisfies the given relations, we have that G is isomorphic to Z_6.

30. $F \oplus Z_3$ where F is the free group on two letters.

32. There are only five groups of order 8: Z_8 and the quaternions have only one element of order 2; $Z_4 \oplus Z_2$ has 3; $Z_2 \oplus Z_2 \oplus Z_2$ has 7; and D_4 has 5.

CHAPTER 26

Symmetry Groups

1. If T is a distance-preserving function and the distance between points a and b is positive, then the distance between $T(a)$ and $T(b)$ is positive.

2. For any fixed v' in $\mathbf{R}^n$ define
$$T_{v'} : \mathbf{R}^n \to \mathbf{R}^n \text{ by } T_{v'}(v) = v + v'.$$
Then $\{T_{v'} \mid v' \in \mathbf{R}^n\}$ is the set of translations of $\mathbf{R}^n$. Closure and associativity follow from the observation
$$T_{v'} \circ T_{w'} = T_{w'+v'};\ T_0 \text{ is the identity; } (T_{v'})^{-1} = T_{-v'}.$$

3. See Figure 1.5.

4. Use Theorem 7.2.

5. There are rotations of $0°, 120°$ and $240°$ about an axis through the centers of the triangles and a $180°$ rotation through an axis perpendicular to a rectangular base and passing through the center of the rectangular base. This gives 6 rotations. Each of these can be combined with the reflection plane perpendicular to the base and bisecting the base. So, the order is 12.

6. 16

7. There are n rotations about an axis through the centers of the n-gons and a $180°$ rotation through an axis perpendicular to a rectangular base and passing through the center of the rectangular base. This gives $2n$ rotations. Each of these can be combined with the reflection plane perpendicular to a rectangular base and bisecting the base. So, the order is $4n$.

8. A drawing or model reveals the group consists of the identity, three $180°$ rotations and 4 reflections and is Abelian.

9. In $\mathbf{R}^1$, there is the identity and an inversion through the center of the segment. In $\mathbf{R}^2$, there are rotations of $0°$ and $180°$, a reflection across the horizontal line containing the segment, and a reflection across the perpendicular bisector of the segment. In $\mathbf{R}^3$, the symmetry group is $G \oplus Z_2$, where G is the plane symmetry group of a circle. (Think of a sphere with the line segment as a diameter. Then G includes any rotation of that sphere about the diameter and any plane containing the diameter of the sphere is a symmetry in G. The Z_2 must be included because there is also an inversion.)

10. No symmetry; symmetry across a horizontal axis only; symmetry across a vertical axis only; symmetry across a horizontal axis and a vertical axis.

11. There are 6 elements of order 4 since for each of the three pairs of opposite squares there are rotations of $90°$ and $270°$.

12. It is the same as a $180°$ rotation.

13. An inversion in $\mathbf{R}^3$ leaves only a single point fixed, while a rotation leaves a line fixed.

14. A rotation of $180°$ about the line L.

15. In $\mathbf{R}^4$, a plane is fixed. In $\mathbf{R}^n$, a hyperplane of dimension $n - 2$ is fixed.

16. Consider a triangle whose sides have lengths a, b, c. The image of this triangle is also a triangle whose sides have lengths a, b, c. Thus the two triangles are congruent (side-side-side).

17. Let T be an isometry, let p, q, and r be the three noncollinear points, and let s be any other point in the plane. Then the quadrilateral determined by $T(p)$, $T(q)$, $T(r)$, and $T(s)$ is congruent to the one formed by p, q, r, and s. Thus, $T(s)$ is uniquely determined by $T(p)$, $T(q)$, and $T(r)$.

18. Use Exercise 17.

19. The only isometry of a plane that fixes exactly one point is a rotation.

20. A translation a distance twice that between a and b along the line joining a and b.

CHAPTER 27

Symmetry and Counting

1. The symmetry group is D_4. Since we have two choices for each vertex, the identity fixes 16 colorings. For R_{90} and R_{270} to fix a coloring, all four corners must have the same color so each of these fixes 2 colorings. For R_{180} to fix a coloring, diagonally opposite vertices must have the same color. So, we have 2 independent choices for coloring the vertices and we can choose 2 colors for each. This gives 4 fixed colorings for R_{180}. For H and V, we can color each of the two vertices on one side of the axis of reflection in 2 ways, giving us 4 fixed points for each of these rotations. For D and D', we can color each of the two fixed vertices with 2 colors and then we are forced to color the remaining two the same. So, this gives us 8 choices for each of these two reflections. Thus, the total number of colorings is

$$\frac{1}{8}(16 + 2 \cdot 2 + 4 + 2 \cdot 4 + 2 \cdot 8) = 6.$$

2. 21

3. The symmetry group is D_3. There are $5^3 - 5 = 120$ colorings without regard to equivalence. The rotations of 120° and 240° can fix a coloring only if all three vertices of the triangle are colored the same so they each fix 0 colorings. A particular reflection will fix a coloring provided that fixed vertex is any of the 5 colors and the other two vertices have matching colors. This gives $5 \cdot 4 = 20$ for each of the three reflections. So, the number of colorings is

$$\frac{1}{6}(120 + 0 + 0 + 3 \cdot 20) = 30.$$

4. 92

5. The symmetry group is D_6. The identity fixes all $2^6 = 64$ arrangements. For R_{60} and R_{300}, once we make a choice of a radical for one vertex, all others must use the same radical. So, these two fix 2 arrangements each. For R_{120} and R_{240} to fix an arrangement, every other vertex must have the same radical. So, once we select a radical for one vertex and a radical for an adjacent vertex, we then have no other choices. So we have 2^2 choices for each of these rotations. For R_{180} to fix an arrangement, each vertex must have the same radical as the vertex diagonally opposite it. Thus, there are 2^3 choices for this case. For the 3 reflections whose axes of symmetry join two vertices, we have 2 choices for each fixed vertex and 2 choices for each of the two vertices on the same side of the reflection axis. This gives us 16 choices for each of these 3 reflections. For the 3 reflections whose axes of reflection bisect opposite sides of the hexagon, we have 2 choices for each of the 3 vertices on the same side of the reflection axis. This gives us 8 choices for each of these 3 reflections. So, the total number of arrangements is

$$\frac{1}{12}(64 + 2 \cdot 2 + 2 \cdot 4 + 8 + 3 \cdot 16 + 3 \cdot 8) = 13.$$

6. 9099

7. The symmetry group is D_4. The identity fixes $6 \cdot 5 \cdot 4 \cdot 3 = 360$ colorings. All other symmetries fix 0 colorings because of the restriction that no color be used more than once. So, the number of colorings is $360/8 = 45$.

8. 231

9. The symmetry group is D_{11}. The identity fixes 2^{11} colorings. Each of the other 10 rotations fixes only the two colorings in which the beads are all the same color. (Here we use the fact that 11 is prime. For example, if the rotation $R_{2 \cdot 360/11}$ fixes a coloring, then once we choose a color for one vertex, the rotation forces all other vertices to have that same color because the rotation moves 2 vertices at a time and 2 is a generator of Z_{11}.) For each reflection, we may color the vertex containing the axis of reflection 2 ways and each vertex on the same side of the axis of reflection 2 ways. This gives us 2^6 colorings for each reflection. So, the number of different colorings is
$$\frac{1}{22}(2^{11} + 10 \cdot 2 + 11 \cdot 2^6) = 126.$$

10. 57

11. The symmetry group is Z_6. The identity fixes all n^6 possible colorings. Since the rotations of 60° and 300° fix only the cases where each section is the same color, they each fix n colorings. Rotations of 120° and 240° each fix n^2 colorings since every other section must have the same color. The 180° rotation fixes n^3 colorings, since once we choose colors for three adjacent sections, the colors for the remaining three sections are determined. So, the number is
$$\frac{1}{6}(n^6 + 2 \cdot n + 2 \cdot n^2 + n^3).$$

12. 51

13. The first part is Exercise 13 in Chapter 6. For the second part, observe that in D_4 we have $\phi_{R_0} = \phi_{R_{180}}$.

14. $\gamma_{g_1 g_2}(x) = (g_1 g_2)xH$

15. R_0, R_{180}, H, V act as the identity and R_{90}, R_{270}, D, D' interchange L_1 and L_2. Then the mapping $g \to \gamma_g$ from D_4 to sym(S) is a group homomorphism with kernel $\{R_0, R_{180}, H, V\}$.

CHAPTER 28

Cayley Digraphs of Groups

1. $4 * (b, a)$
2. $3 * ((a, 0), (b, 0)), (a, 0), (e, 1), 3 * (a, 0), (b, 0), 3 * (a, 0), (e, 1)$
3. $(m/2) * \{3 * [(a, 0), (b, 0)], (a, 0), (e, 1), 3 * (a, 0), (b, 0), 3 * (a, 0), (e, 1)\}$
5. a^3b
6. Say we proceed from x to y via the generators $a_1, a_2, \ldots, a_m$ and via the generators $b_1, b_2, \ldots, b_n$. Then $y = xa_1a_2 \cdots a_m = xb_1b_2 \cdots b_n$ so that $a_1a_2 \cdots a_m = b_1b_2 \cdots b_n$.
7. Both yield paths from e to a^3b.
8. $\text{Cay}\{\{(1, 0), (0, 1)\} : Z_4 \oplus Z_2\}$.

10.

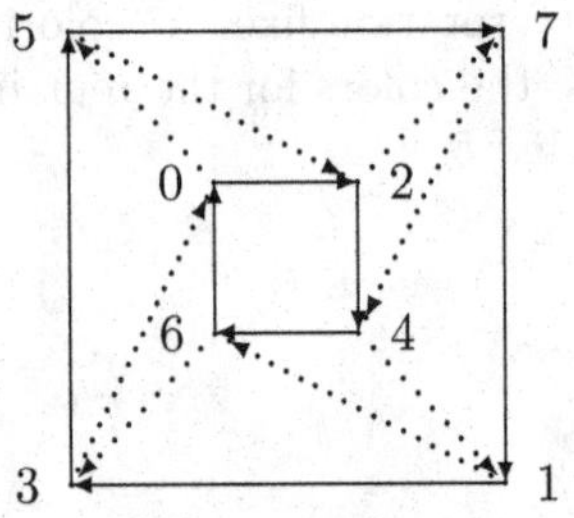

11. Say we start at x. Then we know the vertices $x, xs_1, xs_1s_2, \ldots, xs_1s_2 \cdots s_{n-1}$ are distinct and $x = xs_1s_2 \cdots s_n$. So if we apply the same sequence beginning at y, then cancellation shows that $y, ys_1, ys_1s_2, \ldots, ys_1s_2 \cdots s_{n-1}$ are distinct and $y = ys_1s_2 \cdots s_n$.
12. Trace the sequence b, b, b, a, b, b, b. The digraph could be called undirected because whenever x is connected to y, y is connected to x. Such a digraph (that is, one in which all arrows go both directions) is called a *graph*.
13. If there were a Hamiltonian path from $(0, 0)$ to $(2, 0)$, there would be a Hamiltonian circuit in the digraph, since $(2, 0) + (1, 0) = (0, 0)$. This contradicts Theorem 28.1.
14. $\text{Cay}(\{2, 3\} : Z_6)$ does not have a Hamiltonian circuit.
15. a. If $s_1, s_2, \ldots, s_{n-1}$ traces a Hamiltonian path and $s_is_{i+1} \cdots s_j = e$, then the vertex $s_1s_2 \cdots s_{i-1}$ appears twice. Conversely, if $s_is_{i+1} \cdots s_j \neq e$, then the sequence $e, s_1, s_1s_2, \ldots, s_1s_2 \cdots s_{n-1}$ yields the n vertices (otherwise, cancellation gives a contradiction).

    b. This is immediate from part a.
16. The digraph is the same as those shown in Example 3 except all arrows go in both directions.

17. The sequence traces the digraph in a clockwise fashion.

18. A circuit is $4*((3*a),b)$.

19. Abbreviate $(a,0)$, $(b,0)$, and $(e,1)$ by a, b, and 1, respectively. A circuit is $4*(4*1,a)$, $3*a$, b, $7*a$, 1, b, $3*a$, b, $6*a$, 1, a, b, $3*a$, b, $5*a$, 1, a, a, b, $3*a$, b, $4*a$, 1, $3*a$, b, $3*a$, b, $3*a$, b.

20. Notice that the digraph has four triangles. Start somewhere and call that triangle 1. Now once you enter any of the other three triangles, you must cover all three points before leaving it. The digraph does have a Hamiltonian path, starting at vertex (124) and ending at vertex (1).

21. Abbreviate $(R_{90},0)$, $(H,0)$, and $(R_0,1)$ by R, H, and 1, respectively. A circuit is $3*(R,1,1),H,2*(1,R,R),R,1,R,R,1,H,1,1$.

22. Abbreviate $(a,0)$, $(b,0)$ and $(e,1)$ by a, b and 1 respectively. A circuit is $\frac{m}{2}*(3*(a,b),a,1,3*a,b,3*a,1)$.

23. Abbreviate $(a,0),(b,0)$, and $(e,1)$ by a,b, and 1, respectively. A circuit is $2*(1,1,a),a,b,3*a,1,b,b,a,b,b,1,3*a,b,a,a$.

24. Abbreviate $(a,0)$, $(b,0)$ and $(e,1)$ by a, b and 1 respectively. A circuit is $(m-1)*1,a,2*1,((m-3)/2)*[2*a,b,3*a,1,b,b,a,3*b,1],2*a,b,3*a,1,b,b,a,2*b,1,3*a,b,a,a$.

25. Abbreviate $(r,0),(f,0)$, and $(e,1)$ by r,f, and 1, respectively. Then the sequence is r, r, f, r, r, 1, f, r, r, f, r, 1, r, f, r, r, f, 1, r, r, f, r, r, 1, f, r, r, f, r, 1, r, f, r, r, f, 1.

26. Abbreviate $(r,0)$, $(f,0)$ and $(e,1)$ by r, f and 1 respectively. A circuit in $D_n \oplus Z_{n+1}$ is

$$(n-1)*(n*1,r),n*1,f,n*((n-1)*r,1),(n-1)*r,f.$$

27. $m*((n-1)*(0,1),(1,1))$

28. Adapt the argument given in the proof of Theorem 28.1.

29. Abbreviate $(r,0)$, $(f,0)$, and $(e,1)$ by r,f, and 1, respectively. A circuit is $1,r,1,1,f,r,1,r,1,r,f,1$.

30. Abbreviate $(a,0)$, $(b,0)$ and $(e,1)$ by a,b and 1 respectively. Then a circuit is
$1,a,1,1,b,a,1,a,1,a,b,1,1,$
$2*(a,1,1,b,a,1,a,1,a,b,1,1),a,1,1,b,a,1,a,1,a,b,1$.

31. $5*[3*(1,0),(0,1)],(0,1)]$

32. $12*((1,0),(0,1))$.

33. $12*((1,0),(0,1))$

34.

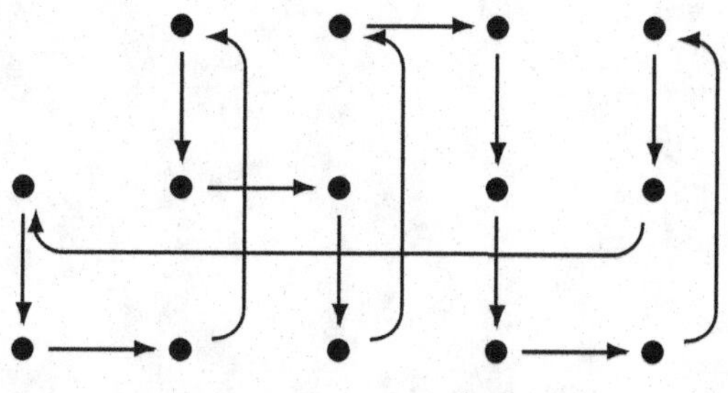

35. Letting V denote a vertical move and H a horizontal move and starting at (1,0), a circuit is $V, V, H, 6 * (V, V, V, H)$.

36. It suffices to show that x travels by a implies xab^{-1} travels by a (for we may successively replace x by xab^{-1}). If xab^{-1} traveled by b, then the vertex xa would appear twice in the circuit.

37. In the proof of Theorem 28.3, we used the hypothesis that G is Abelian in two places: We needed H to satisfy the induction hypothesis, and we needed to form the factor group G/H. Now, if we assume only that G is Hamiltonian, then H also is Hamiltonian and G/H exists.

CHAPTER 29

Introduction to Algebraic Coding Theory

1. wt(000000) = 0; wt(0001011) = 3; wt(0010111) = 4; wt(0100101) = 3; wt(1000110) = 3; wt(1100011) = 4; wt(1010001) = 3; wt(1001101) = 4; etc.

2. 2, 3, 3

3. 1000110; 1110100

4. a. Both $d(u, v)$ and $d(v, u)$ equal the number of positions in which u and v differ.
   b. Use Theorem 29.1.
   c. Use Theorem 29.1.

5. 000000, 100011, 010101, 001110, 110110, 101101, 011011, 111000

6. Argue that $\text{wt}(v) + \text{wt}(u + v) \geq \text{wt}(u)$.

7. By using $t = 1/2$ in the second part of the proof of Theorem 29.2 we have that all single errors can be detected.

8. C' can detect any 3 errors, whereas C can only detect any 2 errors.

9. Observe that a vector has even weight if and only if it can be written as a sum of an even number of vectors of weight 1. So, if u can be written as the sum of $2m$ vectors of even weight and v can be written as the sum of $2n$ vectors of even weight, then $u + v$ can be written as the sum of $2m + 2n$ vectors of even weight and therefore the set of code words of even weight is closed. (We need not check that the inverse of a code word is a code word since every binary code word is its own inverse.)

10. Since the minimum weight of any nonzero member of C is 4, we see by Theorem 29.2 that C will correct any single error and detect any *triple* error. (To verify this, use $t = 3/2$ in the last paragraph of the proof for Theorem 29.2.)

11. No, by Theorem 29.3.

12. $H = \begin{bmatrix} 2 & 2 \\ 1 & 2 \\ 1 & 0 \\ 0 & 1 \end{bmatrix}$

    The code is $\{0000, 1011, 2022, 0121, 0212, 1102, 2201, 2110, 1220\}$. It will correct any single error and detect any double error. 2201.

13. 0000000, 1000111, 0100101, 0010110, 0001011, 1100010, 1010001, 1001100, 0110011, 0101110, 0011101, 1110100, 1101001, 1011010, 0111000, 1111111.

$$H = \begin{bmatrix} 1 & 1 & 1 \\ 1 & 0 & 1 \\ 1 & 1 & 0 \\ 0 & 1 & 1 \\ 1 & 0 & 0 \\ 0 & 1 & 0 \\ 0 & 0 & 1 \end{bmatrix}$$

Yes, the code will detect any single error because it has weight 3.

14. Observe that the subset of code words that end with 0 is a subgroup H. If H is a proper subgroup, note that it has index 2. The same is true for every component.

15. Suppose u is decoded as v, and x is the coset leader of the row containing u. Coset decoding means v is at the head of the column containing u. So, $x + v = u$ and $x = u - v$. Now suppose $u - v$ is a coset leader and u is decoded as y. Then y is at the head of the column containing u. Since v is a code word, $u = u - v + v$ is in the row containing $u - v$. Thus $u - v + y = u$ and $y = v$.

16. For 11101 we get 11100 or 11001. For 01100 we get 11100. No, because the code word could have been 11100 or 11001. Yes, only the code word 11100 differs in one position from the received word.

17. 000000, 100110, 010011, 001101, 110101, 101011, 011110, 111000

$$H = \begin{bmatrix} 1 & 1 & 0 \\ 0 & 1 & 1 \\ 1 & 0 & 1 \\ 1 & 0 & 0 \\ 0 & 1 & 0 \\ 0 & 0 & 1 \end{bmatrix}$$

001001 is decoded as 001101 by all four methods.
011000 is decoded as 111000 by all four methods.
000110 is decoded as 100110 by all four methods.
Since there are no code words whose distance from 100001 is 1 and three whose distance is 2, the nearest-neighbor method will not decode or will arbitrarily choose a code word; parity-check matrix decoding does not decode 100001; the standard-array and syndrome methods decode 100001 as 000000, 110101, or 101011, depending on which of 100001, 010100, or 001010 is a coset leader.

18. Here $2t + s + 1 = 6$. For $t = 0$ and $s = 5$, we can detect any 5 or fewer errors; for $t = 1$ and $s = 3$, we can correct any one error and detect any 2, 3 or 4 errors; for $t = 2$ and $s = 1$, we can correct any 1 or 2 errors and detect any 3 errors.

19. For any received word w, there are only eight possibilities for wH. But each of these eight possibilities satisfies condition 2 or the first portion of condition 3′ of the decoding procedure, so decoding assumes that no error was made or one error was made.

20. The last row is obtained by adding 10000 to each code word. So the code words can be obtained by subtracting 10000 from each member of the last row. (Since the code is binary, this is the same as adding 10000 to each member of the last row.)

21. There are 3^4 code words and 3^6 possible received words.

22. Yes, because the rows are nonzero and distinct.

23. No; row 3 is twice row 1.

24. Suppose that we can use the nearest-neighbor method to correct any t or fewer errors and the weight of the code is $k < 2t + 1$. Let u be a code word of weight k. Let u' be the vector obtained from u by changing $\lceil k/2 \rceil \leq t$ components of u to 0. If

k is even, we have $d(u, u') = \frac{k}{2} = d(0, u')$ so that the nearest neighbor of u' is not unique. If k is odd, then $d(0, u') < d(u, u')$ and u' is not decoded as u.

Now suppose that the nearest-neighbor method will detect any $2t$ or fewer errors and that the weight of the code is at most $2t$. Let u be a code word whose weight is the weight of the code. Then the error made by changing all the components of u to 0 is not detected.

25. No. For if so, nonzero code words would be all words with weight at least 5. But this set is not closed under addition.

26. Say $G = \begin{bmatrix} 1 & 0 & a_1 & a_2 & a_3 \\ 0 & 1 & b_1 & b_2 & b_3 \end{bmatrix}$. To detect 3 errors the minimum weight of nonzero code words must be 4. Thus any nonzero code word has at most one zero component. Since $(10)G = 10a_1a_2a_3$ and $(01)G = 01b_1b_2b_3$ we have $a_i \neq 0$ and $b_i \neq 0$ for $i = 1, 2, 3$. Because $(21)G = 2, 1, 2a_1 + b_1, 2a_2 + b_2, 2a_3 + b_3$ we must have $a_i \neq b_i$. Thus the last three columns for G are $\begin{matrix} 1 \\ 2 \end{matrix}$ or $\begin{matrix} 2 \\ 1 \end{matrix}$. But then $(11)G = 11000$, a contradiction.

27. By Exercise 24, for a linear code to correct every error the minimum weight must be at least 3. Since a (4,2) binary linear code only has three nonzero code words, if each must have weight at least 3, then the only possibilities are (1,1,1,0), (1,1,0,1), (1,0,1,1),(0,1,1,1) and (1,1,1,1). But each pair of these has at least two components that agree. So, the sum of any distinct two of them is a nonzero word of weight at most 2. This contradicts the closure property.

28. 000010 110110 011000 111011 101100 001111 100001 010101.

29. Abbreviate the coset $a + \langle x^2 + x + 1 \rangle$ with a. The following generating matrix will produce the desired code:

$$\begin{bmatrix} 1 & 0 & 1 & 1 & x \\ 0 & 1 & x & x+1 & x+1 \end{bmatrix}.$$

30. $G = \begin{bmatrix} 1 & 0 & 2 & 1 \\ 0 & 1 & 1 & 2 \end{bmatrix}$; $\{0000, 1021, 2012, 0112, 1100, 2121, 0221, 1212, 2200\}$;

$H = \begin{bmatrix} 1 & 2 \\ 2 & 1 \\ 1 & 0 \\ 0 & 1 \end{bmatrix}$. The code will not detect all single errors.

31. By Exercise 14 and the assumption, for each component exactly $n/2$ of the code words have the entry 1. So, determining the sum of the weights of all code words by summing over the contributions made by each component, we obtain $n(n/2)$. Thus, the average weight of a code word is $n/2$.

32. Suppose every vector of weight $t + 1$ is a coset leader. Let v be a code word of weight $2t + 1$ and w the vector obtained from v by changing the first $t + 1$ nonzero component to 0. Then $\text{wt}(w - v) = t + 1$ so that $w - v$ is a coset leader. But $w + C = w - v + C$ and w has weight t. This contradicts the definition of coset leader.

33. Let $c, c' \in C$. Then, $c + (v + c') = v + c + c' \in v + C$ and $(v + c) + (v + c') = c + c' \in C$, so the set $C \cup (v + C)$ is closed under addition.

34. Let v be any vector. If u is a vector of weight 1, then $wt(v)$ and $wt(v+u)$ have opposite parity. Since any vector u of odd weight is the sum of an odd number of vectors of weight 1, it follows that $wt(v)$ and $wt(v+u)$ have opposite parity. Now, mimic the proof of Exercise 23.

35. If the ith component of both u and v is 0, then so is the ith component of $u-v$ and au, where a is a scalar.

CHAPTER 30

Introduction to Galois Theory

1. Note that $\phi(1) = 1$. Thus $\phi(n) = n$. Also, for $n \neq 0$, $1 = \phi(1) = \phi(nn^{-1}) = \phi(n)\phi(n^{-1}) = n\phi(n^{-1})$, so that $1/n = \phi(n^{-1})$. So, by properties of automorphisms, $\phi(m/n) = \phi(mn^{-1}) = \phi(m)\phi(n^{-1}) = \phi(m)\phi(n)^{-1} = mn^{-1} = m/n$.

2. Z_2

3. If α and β are automorphisms that fix F, then $\alpha\beta$ is an automorphism and, for any x in F, we have $(\alpha\beta)(x) = \alpha(\beta(x)) = \alpha(x) = x$. Also, $\alpha(x) = x$ implies, by definition of an inverse function, that $\alpha^{-1}(x) = x$. So, by the Two-Step Subgroup Test, the set is a group.

4. Instead, observe that $Z_2 \oplus Z_2 \oplus Z_2$ has 7 subgroups of order 2.

5. Suppose that a and b are fixed by every element of H. By Exercise 29 in Chapter 13, it suffices to show that $a - b$ and ab^{-1} are fixed by every element of H. By properties of automorphisms we have for any element ϕ of H, $\phi(a-b) = \phi(a) + \phi(-b) = \phi(a) - \phi(b) = a - b$. Also, $\phi(ab^{-1}) = \phi(a)\phi(b^{-1}) = \phi(a)\phi(b)^{-1} = ab^{-1}$.

6. By Exercise 4 of Chapter 19, the splitting field is $Q(\sqrt{2}, i)$. Since $[Q(\sqrt{2}, i) : Q] = 4$, $|\text{Gal}(E/Q)| = 4$. It follows that $\text{Gal}(E/Q) = \{\epsilon, \alpha, \beta, \alpha\beta\}$ where β has fixed field $Q(\sqrt{2})$, α has fixed field $Q(i)$, and $\alpha\beta$ has fixed field $Q(\sqrt{-2})$.

7. It suffices to show that each member of $\text{Gal}(K/F)$ defines a permutation on the a_i's. Let $\alpha \in \text{Gal}(K/F)$ and write $f(x) = c_n x^n + c_{n-1}x^{n-1} + \cdots + c_0$. Then $0 = f(a_i) = c_n a_i^n + c_{n-1}a_i^{n-1} + \cdots + c_0$. So, $0 = \alpha(0) = \alpha(c_n)(\alpha(a_i))^n + \alpha(c_{n-1})\alpha(a_i)^{n-1} + \cdots + \alpha(c_0) = c_n(\alpha(a_i))^n + c_{n-1}\alpha(a_i)^{n-1} + \cdots + c_0 = f(\alpha(a_i))$. So, $\alpha(a_i) = a_j$ for some j, and therefore α permutes the a_i's.

8. Use Corollary 3 of Theorem 16.2 and Exercise 7 of this chapter.

9. Observe that $\phi^6(\omega) = \omega^{729} = \omega$, whereas $\phi^3(\omega) = \omega^{27} = \omega^{-1}$ and $\phi^2(\omega) = \omega^9 = \omega^2$. $\phi^3(\omega + \omega^{-1}) = \omega^{27} + \omega^{-27} = \omega^{-1} + \omega$. $\phi^2(\omega^3 + \omega^5 + \omega^6) = \omega^{27} + \omega^{45} + \omega^{54} = \omega^6 + \omega^3 + \omega^5$.

10. $|\text{Gal}(E/Q)| = [E : Q] = 4$; $|\text{Gal}(Q(\sqrt{10})/Q| = [Q(\sqrt{10}) : Q] = 2$.

11. a. $Z_{20} \oplus Z_2$ has three subgroups of order 10. b. 25 does not divide 40 so there is none. c. $Z_{20} \oplus Z_2$ has one subgroup of order 5.

12. See Example 4 in this chapter.

13. The splitting field over $\mathbf{R}$ is $\mathbf{R}(\sqrt{-3})$. The Galois group is the identity and the mapping $a + b\sqrt{-3} \to a - b\sqrt{-3}$.

14. Observe that D_6 has exactly three subgroups of order 6.

15. Use Theorem 21.3.

16. Use the Corollary to Theorem 23.2 and Theorem 11.1.

17. If there were a subfield K of E such that $[K : F] = 2$ then, by the Fundamental Theorem of Galois Theory (Theorem 30.1), A_4 would have a subgroup of index 2. But by Example 5 in Chapter 7, A_4 has no such subgroup.

18. Let $\omega = (-1 + i\sqrt{3})/2$. The splitting field of $x^3 - 1$ over Q is $Q(\omega)$. Since $[Q(\omega) : Q] = 2$, the Galois group of $x^3 - 1$ over Q is Z_2. The splitting field of $x^3 - 2$ over Q is $Q(\sqrt[3]{2}, \omega)$. Since $[Q(\sqrt[3]{2}, \omega) : Q] = 6$, the Galois group has order 6 and is generated by α and β where $\alpha(\sqrt[3]{2}) = \omega\sqrt[3]{2}$, $\alpha(\omega) = \omega$ and $\beta(\sqrt[3]{2}) = \sqrt[3]{2}$, $\beta(\omega) = \omega^2$. Since $\alpha\beta \neq \beta\alpha$, the group must be S_3 (see Theorem 7.2 and the remark at the end of the proof).

19. This follows directly from the Fundamental Theorem of Galois Theory (Theorem 30.1) and Sylow's First Theorem (Theorem 23.3).

20. Use the subgroup lattice for D_5.

21. Let ω be a primitive cube root of 1. Then $Q \subset Q(\sqrt[3]{2}) \subset Q(\omega, \sqrt[3]{2})$ and $Q(\sqrt[3]{2})$ is not the splitting field of a polynomial in $Q[x]$.

22. Use the Fundamental Theorem and the fact that $\mathrm{Gal}(E/F)$ is finite.

23. By the Fundamental Theorem of Finite Abelian Groups (Theorem 11.1), the only Abelian group of order 10 is Z_{10}. By the Fundamental Theorem of Cyclic Groups (Theorem 4.3), the only proper, nontrivial subgroups of Z_{10} are one of index 2 and one of index 5. So, the lattice of subgroups of Z_{10} is a diamond with Z_{10} at the top, $\{0\}$ at the bottom, and the subgroups of indexes 2 and 5 in the middle layer. Then, by the Fundamental Theorem of Galois Theory, the lattice of subfields between E and F is a diamond with subfields of indexes 2 and 5 in the middle layer.

24. $Q(\omega + \omega^4)$

25. By Example 7, the group is Z_6.

26. Z_3

27. This follows directly from Exercise 21 in Chapter 24.

28. Let K be the subgroup of rotations in D_n. The desired series is $\{R_0\} \subset K \subset D_n$.

29. This follows directly from Exercise 43 in Chapter 23.

30. Note that A_4 has a normal Sylow 2-subgroup.

31. This follows directly from Exercise 50 in Chapter 10.

32. Let $\{e\} = H_0 \subset H_1 \subset \cdots \subset H_n = G$ be the series that shows that G is solvable. Then $H_0 \cap H \subset H_1 \cap H \subset \cdots \subset H_n \cap H = H$ shows that H is solvable.

33. Since $K/N \lhd G/N$, for any $x \in G$ and $k \in K$, there is a $k' \in K$ such that $k'N = (xN)(kN)(xN)^{-1} = xNkNx^{-1}N = xkx^{-1}N$. So, $xkx^{-1} = k'n$ for some $n \in N$. And since $N \subseteq K$, we have $k'n \in K$.

34. Use parts 7, 6 and 1 of Theorem 15.1.

35. Since G is solvable there is a series

$$\{e\} = K_0 \subset K_1 \subset \cdots \subset K_m = G$$

such that K_{i+1}/K_i is Abelian. Now there is a series

$$\frac{K_i}{K_i} = \frac{L_0}{K_i} \subset \frac{L_1}{K_i} \subset \cdots \subset \frac{L_t}{K_i} = \frac{K_{i+1}}{K_i},$$

where $|(L_{j+1}/K_i)/(L_j/K_i)|$ is prime. Then

$$K_i = L_0 \subset L_1 \subset L_2 \subset \cdots \subset L_t = K_{i+1}$$

and each $|L_{j+1}/L_j|$ is prime (see Exercise 42 of Chapter 10). We may repeat this process for each i.

36. Mimic the analysis carried out for $3x^5 - 15x + 5$ at the end of Chapter 30.

CHAPTER 31

Cyclotomic Extensions

1. Since $\omega = \cos\frac{\pi}{3} + i\sin\frac{\pi}{3} = \cos\frac{2\pi}{6} + i\sin\frac{2\pi}{6}$, ω is a zero of $x^6 - 1 = \Phi_1(x)\Phi_2(x)\Phi_3(x)\Phi_6(x) = (x-1)(x+1))(x^2+x+1)(x^2-x+1)$, it follows that the minimal polynomial for ω over Q is $x^2 - x + 1$.

2. Use Theorem 31.1

3. Over Z, $x^8 - 1 = (x-1)(x+1)(x^2+1)(x^4+1)$. Over Z_2, $x^2+1 = (x+1)^2$ and $x^4 + 1 = (x+1)^4$. So, over Z_2, $x^8 - 1 = (x+1)^8$. Over Z_3, x^2+1 is irreducible, but x^4+1 factors into irreducibles as $(x^2+x+2)(x^2-x-1)$. So, $x^8 - 1 = (x-1)(x+1)(x^2+1)(x^2+x+2)(x^2-x-1)$. Over Z_5, $x^2+1 = (x-2)(x+2), x^4+1 = (x^2+2)(x^2-2)$, and these last two factors are irreducible. So, $x^8-1 = (x-1)(x+1)(x-2)(x+2)(x^2+2)(x^2-2)$.

4. Use $x^n - 1 = (x-1)(x^{n-1} + x^{n-2} + \cdots + x + 1)$ and that fact that nth roots of unity form a cyclic group of order n.

5. Let ω be a primitive nth root of unity. We must prove $\omega\omega^2\cdots\omega^n = (-1)^{n+1}$. Observe that $\omega\omega^2\cdots\omega^n = \omega^{n(n+1)/2}$. When n is odd, $\omega^{n(n+1)/2} = (\omega^n)^{(n+1)/2} = 1^{(n+1)/2} = 1$. When n is even, $(\omega^{n/2})^{n+1} = (-1)^{n+1} = -1$.

6. $\Phi_3(x)$.

7. If $[F:Q] = n$ and F has infinitely many roots of unity, then there is no finite bound on their multiplicative orders. Let ω be a primitive mth root of unity in F such that $\phi(m) > n$. Then $[Q(\omega):Q] = \phi(m)$. But $F \supseteq Q(\omega) \supseteq Q$ implies $[Q(\omega):Q] \leq n$.

8. Observe that $x^n - 1 = (x-1)(x^{n-1} + x^{n-2} + \cdots + x + 1)$ and use Theorem 31.1.

9. Let $2^n + 1 = q$. Then $2 \in U(q)$ and $2^n = q - 1 = -1$ in $U(q)$ implies that $|2| = 2n$. So, by Lagrange's Theorem, $2n$ divides $|U(q)| = q - 1 = 2^n$.

10. We know $\Phi_2(0) = 1$. Now observe that
$$x^n - 1 = (x-1)\prod_{\substack{d|n\\1<d<n}} \Phi_d(x)\Phi_n(x)$$
and use induction.

11. Let ω be a primitive nth root of unity. Then $2n$th roots of unity are $\pm 1, \pm\omega, \ldots, \pm\omega^{n-1}$. These are distinct, since $-1 = (-\omega^i)^n$, whereas $1 = (\omega^i)^n$.

12. Let α be a primitive mnth root of unity. Then α^n is a primitive mth root of unity and α^m is a primitive nth root of unity. This shows that $(x^m-1)(x^n-1)$ splits in the splitting field of $x^{mn} - 1$. Conversely, let β be a primitive mth root of unity and γ be a primitive nth root of unity. It suffices to show that $|\beta\gamma| = mn$. Let $H = \langle\beta\gamma\rangle$. Since $(\beta\gamma)^{mn} = (\beta^m)^n(\gamma^n)^m = 1 \cdot 1$ we know $|H| \leq mn$. Since $(\beta\gamma)^m = \beta^m\gamma^n = \gamma^n$ and, by Theorem 4.2, $|\gamma^n| = |\gamma|$, we know that n divides $|H|$. By symmetry, m divides $|H|$. Thus $|H| \geq mn$. This proves that the splitting field of $(x^m-1)(x^n-1)$ contains a primitive mnth root of unity.

13. First observe that deg $\Phi_{2n}(x) = \phi(2n) = \phi(n)$ and deg $\Phi_n(-x) =$ deg $\Phi_n(x) = \phi(n)$. Thus, it suffices to show that every zero of $\Phi_n(-x)$ is a zero of $\Phi_{2n}(x)$. But ω is a zero of $\Phi_n(-x)$ means that $|-\omega| = n$, which in turn implies that $|\omega| = 2n$. (Here $|\omega|$ means the order of the group element ω.)

14. Since the two sides are monic and have the same degree it suffices to prove that every zero of $\Phi_{p^k}(x)$ is a zero of $\Phi_p(x^{p^{k-1}})$. Let ω be a zero of $\Phi_{p^k}(x)$ and note that $|\omega| = p^k$ implies that $|\omega^{p^{k-1}}| = p$.
$\Phi_8(x) = x^4 + 1, \Phi_{27}(x) = (x^9)^2 + x^9 + 1 = x^{18} + x^9 + 1$.

15. Let $G = \mathrm{Gal}(Q(\omega)/Q)$ and H_1 be the subgroup of G of order 2 that fixes $\cos(\frac{2\pi}{n})$. Then, by induction, G/H_1 has a series of subgroups
$H_1/H_1 \subset H_2/H_1 \subset \cdots \subset H_t/H_1 = G/H_1$, so that $|H_{i+1}/H_1 : H_i/H_1| = 2$. Now observe that $|H_{i+1}/H_1 : H_i/H_1| = |H_{i+1}/H_i|$.

16. Use Theorem 31.4.

17. Instead, we prove that $\Phi_n(x)\Phi_{pn}(x) = \Phi_n(x^p)$. Since both sides are monic and have degree $p\phi(n)$, it suffices to show that every zero of $\Phi_n(x)\Phi_{pn}(x)$ is a zero of $\Phi_n(x^p)$. If ω is a zero of $\Phi_n(x)$, then $|\omega| = n$. By Theorem 4.2, $|\omega^p| = n$ also. Thus ω is a zero of $\Phi_n(x^p)$. If ω is a zero of $\Phi_{np}(x)$, then $|\omega| = np$ and therefore $|\omega^p| = n$.

18. Use Theorem 31.4.

19. Let ω be a primitive 5th root of unity. Then the splitting field for $x^5 - 1$ over Q is $Q(\omega)$. By Theorem 31.4, $\mathrm{Gal}(Q(\omega)/Q) \approx U(5) \approx Z_4$. Since $\langle 2 \rangle$ is the unique subgroup strictly between $\{0\}$ and Z_4, we know by Theorem 32.1 that there is a unique subfield strictly between Q and E.

20. Use Theorem 31.4 and Theorem 30.1.

21. Suppose that a prime $p = 2^m + 1$ and m is not a power of 2. Then $m = st$ where s is an odd integer greater than 1 (the case where $m = 1$ is trivial). Let $n = 2^t + 1$. Then $1 < n < p$ and $2^t \bmod n = -1$. Now looking at $p \bmod n$ and replacing 2^t with -1, we have $(2^t)^s + 1 = (-1)^s + 1 = 0$. This means that n divides the prime p, which is a contradiction.

22. The three automorphisms that take $\omega \to \omega^4, \omega \to \omega^{-1}, \omega \to \omega^{-4}$ have order 2.

13. [illegible] the degree [illegible] and let Φ_n [illegible]. Then [illegible] of Φ_n [illegible] of degree [illegible] Galois group [illegible] equals the order of the group [illegible].

[illegible] the [illegible] roots [illegible].

[illegible] $\mathbb{Q}$ [illegible]. The group [illegible] C_2 [illegible] subgroups [illegible] H_1 [illegible] H_2 [illegible].

15. Use Theorem [illegible].

17. In [illegible] we prove [illegible]. Since both sides are monic and have degree [illegible] it suffices to show that every zero of [illegible] is a zero of [illegible]. [illegible] is a zero of [illegible] then [illegible] by Theorem [illegible]. [illegible] zero of $\Phi_n(x^p)$ [illegible].

19. Use Theorem [illegible].

[illegible] is a primitive [illegible] root of unity. Then the splitting field for [illegible] is [illegible]. The Galois group [illegible] $\mathrm{Gal}(\mathbb{Q}(\zeta)/\mathbb{Q})$ [illegible] $\mathbb{Z}_2$ [illegible] is the unique subgroup [illegible] between $\mathbb{Q}$ and [illegible], we know by Theorem [illegible] that there is a unique subfield [illegible] between $\mathbb{Q}$ and [illegible].

20. Use Theorem [illegible] and Theorem [illegible].

21. Suppose that a prime p [illegible] and [illegible] power of 2. [illegible] where [illegible] is trivial. Let [illegible] Then [illegible] and [illegible] with [illegible] is the [illegible].

22. The three automorphisms that [illegible] have order 2.

Printed in the United States
by Baker & Taylor Publisher Services

Printed in the United States
by Baker & Taylor Publisher Services